T-Labs Series in Telecommunication Services

It is the aim of the Springer Series in Telecommunication Services to foster an interdisciplinary exchange of knowledge addressing all topics which are essential for developing high-quality and highly usable telecommunication services. This includes basic concepts of underlying technologies, distribution networks, architectures and platforms for service design, deployment and adaptation, as well as the users' perception of telecommunication services. By taking a vertical perspective over all these steps, we aim to provide the scientific bases for the development and continuous evaluation of innovative services which provide a better value for their users. In fact, the human-centric design of high-quality telecommunication services – the so called "quality engineering" – forms an essential topic of this series, as it will ultimately lead to better user experience and acceptance. The series is directed towards both scientists and practitioners from all related disciplines and industries.

** Indexing: books in this series are indexing in Scopus **

Tanja Kojić

User Experience for Serious Games in Virtual Reality

Springer

Deutsche Telekom Laboratories

Tanja Kojić
Quality and Usability Lab
TU Berlin
Berlin, Berlin, Germany

ISSN 2192-2810 ISSN 2192-2829 (electronic)
T-Labs Series in Telecommunication Services
ISBN 978-3-031-75532-3 ISBN 978-3-031-75530-9 (eBook)
https://doi.org/10.1007/978-3-031-75530-9

This Springer imprint is published by the registered company Springer Nature Switzerland AG
The registered company address is: Gewerbestrasse 11, 6330 Cham, Switzerland

Preface

Although Virtual Reality (VR) technology has been part of research for some time, several VR devices have recently been introduced to the market. The prediction for VR is to become even more popular in different sectors, such as VR serious gaming, including the contexts outside of just entertainment. One method for generating content for VR technology is to mimic already-defined design criteria for online and video games. That, however, is not the best option since techniques for producing VR content should allow an immersive narrative to improve realism and enjoyment.

The primary purpose of this work is to evaluate and identify different influences on User Experience (UX) for VR serious games, intending to narrow down the research gap between Influencing Factors (IFs), UX, and design guidelines for VR serious games. With eight user studies and five different VR serious games developed, different influences and links between those factors and UX components are investigated. The focus of this work is on commercially accessible VR equipment, particularly Head Mounted Display (HMD), and questionnaires as a research method used for collecting data.

Interesting results have been found for all Influencing Factors (IFs) categories: System with Content, Context, and Human. According to the findings, using a virtual environment in VR increases sensations of general presence, spatial presence, and perceived realism. Furthermore, studies that have measured the effect of content and virtual surroundings demonstrate how various activities affected users to prefer different settings for the VR serious game. When it comes to context, different social situations impact users' moods and sense of presence, as well as their overall perceptions of the UX and almost all characteristics of social acceptability. The findings show that even when just transferring from having one person to a few people being near the user, the acceptability of VR technology in public areas is dramatically decreased. Also, when it comes to user demographics and the influence of human factors, age and gender have been shown to impact parts of UX for VR serious gaming, and people with different prior experience with VR systems are likely to have different set-up preferences.

This work reports all connections between IFs and particular subfactors to UX components where significant effects were found as a first step to creating a full UX model for VR serious games. Finally, the presented work provides the means for ready-to-use proposed design guidelines formed on results and studied examples of VR serious gaming applications.

Berlin, Germany Tanja Kojić

Acknowledgments

First, I would like to offer my most sincere thankfulness to my supervisor, Prof. Dr.-Ing. Sebastian Moller, for giving me continuous support over the course of all of these years. I would really like to express how much I appreciate your constant encouragement, time, effort, and all of opportunities.

Special thanks go to Prof. Dr.-Ing. Jan-Niklas Voigt-Antons, from the first day, I have always considered you to be as well my mentor with this work. Thank you for always being there, for small and big tasks, for listening to my smart and crazy ideas, and for dealing with on-time and last-minute deadlines. I am very thankful for all of your help and support.

Next, I would like to thank my colleagues whom not only have I shared my work with but everyday stories, joy, and inspirations. The great energy in this team made sailing through a working day a lot easier on any given day. Thank you: Maurizio Vergari, Kerstin Pieper, Robert P. Spang, Thilo Michael, Dr.-Ing. Steven Schmidt, Martin Josef Burghart, Sonia Sobol, Danish Ali, Neslihan Iskender, Vera Schmitt, Britta Hesse, Wafaa Wardah, Salar Mohtaj, Sai Sirisha Rallabandi, Luis Meier, Dr.-Ing. Stefan Uhrig, Dr.-Ing. Gabriel Mittag, Dr.-Ing. Rafael Zequeira Jiménez, Dr. Andres Pinilla Palacios, Dr.-Ing. Saman Zadtootaghaj, Dr.-Ing. Tim Polzehl, Dr.-Ing. Babak Naderi, and Dr.-Ing. Stefan Hillmann.

As well, special thank you to Irene Hube-Achter and Yasmin Hillebrenner for your amazing administrative work, your endless support, and your energy to find answers and solutions for each and every administrative situation. Sometimes your solutions feel like magic, as we don't even know how it works—but it does every time. Thank you also to Tobias Jettkowski for your technical support with equipment and arrangements for this work.

I would also like to thank the members of my committee, Assoc. Prof. Dr. Katrien De Moor and Asst. Prof. Dr. Sc. Mirko Sučnjević, I really appreciate your time and work in co-reviewing my dissertation.

Finally, I would like to express my gratitude to my beloved family and friends. Without your support and encouragement, this journey would not have been possible.

Contents

List of Abbreviations

ANOVA	Analysis of Variance. 79
AR	Augmented Reality. 12
ATI	Affinity for Technology Interaction. 74
AV	Augmented Virtuality. 12
CAVE	Cave Automatic Virtual Environment. 14
DTC	dyadic telephone conversation. 58
FSS	Flow State Scale. 26
GOMS	Goals, Operators, Methods, and Selection Rules. 23
HCI	Human-Computer Interaction. 21
HMD	Head Mounted Display. 14
HUD	Heads-Up Display. 48
IF	Influencing Factor. 31
IPQ	Igroup Presence Questionnaire. 27
ISO	International Organization for Standardization. 21
ITU	International Telecommunication Union. 4
IVE	Immersive Virtual Environment. 31
MPAM	Motives for Physical Activity Examine - Revised. 26
MR	Mixed Reality. 12
NMQ	Networked Minds Measure of Social Presence. 58
PC	Pair Comparison. 47
PQ	Presence Questionnaire. 27
QoE	Quality of Experience. 22
RQ	Research Question. 4
SAM	Self-Assessment-Manikin. 24
SAQ	Social Acceptability Questionnaire. 30
SDT	Self-Determination Theory. 25
SMS	Sport Motivation Scale. 26

SSE	Simulated Social Environment. 55
SUS	System Usability Scale. 23
TV	television. 58
UEQ	User Experience Questionnaire. 24
UI	user interface. 15
USE	Usefulness, Satisfaction, and Ease of use. 23
UX	User Experience. 2
UXIVE	User Experience in Immersive Virtual Environment. 22
VE	Virtual Environment. 3
VoIP	Voice over Internet Protocol. 49
VR	Virtual Reality. 12
WCAG2.1	Web Content Accessibility Guidelines 2.1. 29
WMA	World Medical Association. 77

Chapter 1
Introduction

1.1 Motivation

Great deal of progress has been made in the last few years with Virtual Reality (VR) technology [1] as VR headsets of different complexity are now available to the general public and are used for more than just games. Today's use cases for VR are many, owing to the ability to generate and immerse people in a variety of different virtual worlds and environments [2]. These virtual worlds replicate or even enhance the real world [3], hence expanding the prospects and market for VR. Each year, the number of individuals purchasing immersive technology grows, and by the end of 2024, it is estimated that VR will generate around 12.19 billion dollars in revenue worldwide [4]. Further, predictions for the foreseeable future imply that such numbers will only rise. Even though most of its current popularity can be connected to the VR gaming industry, still, VR technology can be found and used in many other aspects. It can be used for a range of fields [5], including but not limited to education, sports, training, tourism, simulators, big data visualization, health care issues, and more. As a result, the purpose of the VR application in those fields is not the game itself but rather including game-design principles in a nongame context [6]. This means it is not primarily intended for entertainment, leading to those games being referred to as serious games [7].

At the same time, while VR technology is gaining popularity, researchers have also been focusing on video gaming. They have determined that German teenagers spend, on average, 117 minutes per day playing video games [8], concluding that gaming is becoming a significant part of many people's lives in recent years. Furthermore, the intrinsic motivation that players are driven by is not limited only to young people; most players engage in their activity without receiving any external benefits. This phenomenon has recently become interesting for both the industry and research community, which aim to understand why players are so motivated and engaged with their games.

T. Kojić, *User Experience for Serious Games in Virtual Reality*, T-Labs Series in Telecommunication Services, https://doi.org/10.1007/978-3-031-75530-9_1

As a result, the discipline of gamification was established, which studies the use of game components, and game-design methodologies in nongame contexts [6]. The objective of it is to improve user engagement through the usage of games, but also to expand the theory of gamification into having defined games that are produced for a primary non-entertainment goal (serious games) [7].

Since the VR industry is still being established at the moment, despite the fact that VR technology is not new, most of the companies in this market have introduced their first VR devices in the recent few years. Based on these first-generation virtual reality devices, companies are gathering data about usage and forming research departments while attempting to decide the future path of the developing project. So far, researchers have pointed out several challenges about VR technology, such as motion sickness [9], resolution [10], field of view [11], and followingly the design of the User Experience (UX) [12]. One of the approaches to creating content for VR technology is to replicate already established design guidelines for web and video games. However, that is not the best approach, as methodologies and techniques for designing VR content should enable successful immersive storytelling to enhance realism and entertainment.

While developing VR applications, it should be attempted to deliver a highly responsive solution to the user's demands in accordance with the concept of user-centric design [13]. Depending on how successfully this methodology is implemented in the application, the user experience will be better [14]. For optimal user experience, gathering data, including user evaluation, is necessary under various relevant conditions to develop suggestions, standards, and design guidelines. As a result, in order to adequately cover relevant aspects while preparing empirical studies, a broad understanding of influencing factors for VR serious games must be accessible. By the time work on the current dissertation began, the assessment procedures employed in the research community, in particular, were still quite restricted. A standardized set of recommendations and design guidelines for VR user experience would be extremely useful for academics and companies since it would provide a focused overview of influencing factors for VR serious gaming.

Finally, the present research aims to narrow the research gap between the design and user experience for serious games in VR. Therefore special attention is paid to assessing influencing factors for serious game quality, which are a critical component as they form the framework of design guidelines for future research.

1.2 Problem Statement

As VR technologies are rapidly expanding, their technical components are diverse and complex. Furthermore, there is a wide range of possible influence factors and various dimensions that might influence user experiences. As a result, some research regarding virtual reality and user experience had to be reduced in scope.

For the presented research, the focus is on serious games in virtual reality as it is one of the ways for industry organizations to get their messages across in more meaningful, interesting, and engaging ways. Serious games are becoming more important, which means service providers have an economic incentive to improve their service and to optimize the user experience for their customers. This research should enable designers and developers of such VR serious games to enhance the experience for their customers by focusing on how to utilize currently available technical solutions for the best application and system design rather than the improvement on different technology set-ups, networks, or encoding parameters.

Consequently, because the research focuses on market-ready VR technology, it uses the current state of devices, namely VR solutions with a head-mounted display (HMD), whereas alternative set-ups are outside the scope of the research. In addition, technical characteristics such as device resolution, audio settings, network characteristics, and encoding parameters are employed as given by the system, and these features are outside the scope of this research. However, some information about the technological system is crucial because it is a major influencing factor in gaming quality of experience. Still, further enhancements in the near future, such as increased resolution, frame rates, and other technological factors, are likely to enhance the experience.

With regard to testing participants for subjective studies, the physical testing of vision or hearing was not tested but only reported. Similarly, immersion tendency or simulation sickness was just addressed and not further investigated in this research. Physical testing of such factors could be done with some physiological measurements; however, those analyses were out of scope for this research, and participants were only asked to report on their physical state. This was done in such a manner as the majority of current applications save such user characteristics in settings only based on reported values and do not monitor them using any additional sensors.

Lastly, the impact of temporal context elements, such as the frequency and intensity of usage, was not further investigated in this study, and the VR systems were only used within the time period that was supported by the producers. This restriction was incorporated because symptoms of simulator sickness, such as dizziness, loss of spatial awareness, nausea, and eye pain, typically worsen as the amount of time spent using the simulator increases, and the overall experience may be considerably reduced as a result of these consequences. Thus, a time restriction was implemented, and participants were free to report symptoms and withdraw from the trial at any time.

While this research primarily focuses on VR serious games, some of the findings about influencing factors and recommendations based on them may also be applicable to other immersive media fields. In particular, VR gaming, where the primary focus is on entertainment, can also benefit from this research, as many of the factors are similar. Further on, passive video streaming applications where only 360-degree video content is streamed in a Virtual Environment (VE) to the passive spectators are not considered serious games as there is no gaming element, but they might as well benefit from some of the findings. These fields are researched separately and are not included in the scope. However, due to the similarity of the technology, it is to be expected that similar conclusions could be applicable.

Prior work on standards and recommendations for virtual reality, gaming, and factors influencing the quality of experience must be taken into consideration when placing this presented work in the context of the state of the art. It is important to note that the International Telecommunication Union (ITU) Recommendation G.1035: Influencing factors on quality of experience for virtual reality services [15], as well as the QUALINET White Paper on Definitions of Immersive Media Experience (IMEx) [16], which elaborated on the Influencing factors on immersive media experience as well as the assessment of immersive media experience, was used to categorize virtual reality in this manner.

Following the discussion of the motivation and the scope of the work given above, the following section summarizes the thesis's overall aim and the corresponding research questions (RQ).

Aim of Thesis
The main aim of this work is to assess user experience for VR serious games and determine what factors influence it, with the goal to narrow the research gap between influencing factors, user experience, and design guidelines of such applications.
Based on an analysis of several empirically investigated VR serious game scenarios, this research presents an overview of the significant influencing factors for VR serious gaming and the relationship between those factors and user experience components.
Finally, the presented work provides the means for ready-to-use proposed design guidelines formed on results and studied VR serious gaming application examples.

1.3 Prior Publications

The next section describes the author's publications, which serve as the foundation for the work presented, as well as where to find them in the thesis:

- S. Schmidt, P. Ehrenbrink, B. Weiss, J.-N. Voigt-Antons, T. Kojić, A. Johnston and S. Möller, "Impact of Virtual Environments on Motivation and Engagement During Exergames", in *2018 Tenth International Conference on Quality of Multimedia Experience (QoMEX)*, IEEE, 2018, pp. 1–6 [17].

This paper investigates the Quality of Experience of participants using the different VR system set-ups (HMD and CAVE) compared to the system without a VE. Measurements include concepts of flow, presence, and video quality, with the results showing significant advantages of the HMD as well as of the CAVE, but the HMD was favored by the majority of participants due to a superior feeling of presence. This paper was a foundation for selecting systems for this work and is part of Chap. 2. This study was a joint project between Quality and Usability Lab, Technische Universität Berlin, and Creativity and Cognition Studios, University of Technology Sydney. As part of it, Steven Schmidt, Patrick Ehrenbrink, Benjamin Weiss, and Andrew Johnston conducted the study in a laboratory in Sydney. All authors of this work were jointly included in the analysis of results, where Tanja Kojić was responsible for the related work section of the paper.

- T. Kojić, J.-N. Voigt-Antons, S. Schmidt, L. Tetzlaff, B. Kortowski, U. Sirotina and S. Möller, "Influence of Virtual Environments and Conversations on User Engagement During Multiplayer Exergames", in *2018 Tenth International Conference on Quality of Multimedia Experience (QoMEX)*, IEEE, 2018, pp. 1–3 [18].

For this paper, the focus was to investigate the influence of virtual environments and the possibility of conversation on the presence and social presence, where results showed a significant increase in the perceived feeling of presence in VR, leading as well to an increase in user experience. Mentioned effects are included in Chaps. 4 and 6. Regarding the responsibilities, Tanja Kojić was responsible for all essential activities, including research design, data analysis, paper writing, and conducting subjective user studies. All authors were as well included in the latter, while Sebastian Möller and Jan-Niklas Voigt-Antons contributed to the study design itself:

- T. Kojić, U. Sirotina, S. Möller and J.-N. Voigt-Antons, "Influence of UI Complexity and Positioning on User Experience During VR Exergames", in *2019 Eleventh International Conference on Quality of Multimedia Experience (QoMEX)*, IEEE, 2019, pp. 1–6 [19].

This paper's results contribute to Chaps. 5 and 7 as it shows that the level of UI complexity impacts how easy it is to read, and the placement of UI elements impacts how users think the system helps them. Additionally, participants preferred a different level of complexity depending on where the metrics were shown to them.

This paper is based on results gathered for the Master thesis of Uliana Sirotina, where she has conducted the user studies and done the basic analysis of results for it together with Tanja Kojić. Furthermore, Tanja Kojić was responsible for the paper writing process, while the study design was a joint work of all authors:

- T. Kojic, S. Schmidt, S. Möller and J.-N. Voigt-Antons, "Influence of Network Delay in Virtual Reality Multiplayer Exergames: Who is actually delayed?", in 2019 Eleventh International Conference on Quality of Multimedia Experience (QoMEX), IEEE, 2019, pp. 1–3 [20].

This paper presents the different perceptions of delay and QoE depending on a user's own network delay, and these effects are included in Chap. 4. For this publication, Tanja Kojić supervised the user study that was performed by Bruno Kortowski, and she was responsible for the major parts of the paper writing process. Additionally, she was responsible for further analysis of the data together with Steven Schmidt and Jan-Niklas Voigt-Antons, while Sebastian Möller and Jan-

Niklas Voigt-Antons contributed to the study design and the supervision of the project.

- T. Kojić, L. T. Nugyen, and J.-N. Voigt-Antons, "Impact of Constant Visual Biofeedback on User Experience in Virtual Reality Exergames", in *2019 IEEE International Symposium on Multimedia (ISM)*, IEEE, 2019, pp. 307–3073 [21].

The following paper focuses on comparing the effects of several different visualizations of biofeedback, particularly breathing rhythm. Participants rated a solution in VR significantly more sympathetic than one without VR on the sympathy scale. Further on, the flow was also rated statistically significantly better for all instruction types in VR. All of these reported effects are included in Chap. 4 of this thesis. Regarding the contributions in this paper, all authors have contributed to the study design and data analyses, Lan Thao Nugyen performed the user study, and the paper was written by Tanja Kojić.

- T. Kojić, D. Ali, R. Greinacher, S. Möller and J.-N. Voigt-Antons, "User Experience of Reading in Virtual Reality—Finding Values for Text Distance, Size and Contrast", in 2020 Twelfth International Conference on Quality of Multimedia Experience (QoMEX), IEEE, 2020, pp. 1–6 [22].

This paper investigates what values users find pleasant when reading in VR, with the results providing text parameters such as mean values for angular size and color contrast, which vary based on the text's length. These results are mentioned as part of Chap. 5. As part of the research for this paper, Tanja Kojić was responsible for all stages, including research design (together with Jan-Niklas Voigt-Antons and Sebastian Möller), integration of settings (together with Danish Ali and Robert Spang), analysis of the data, conducting subjective user studies and finally paper writing.

- J.-N. Voigt-Antons, T. Kojić, D. Ali and S.Möller, "Influence of Hand Tracking as a Way of Interaction in Virtual Reality on User Experience", in 2020 Twelfth International Conference on Quality of Multimedia Experience (QoMEX), IEEE, 2020, pp. 1–4 [23].

This paper presents research comparing different VR interaction types (using controllers or hand tracking), with findings indicating that, depending on the different hand visualizations and tasks in the application, different interaction types statistically significantly influence reported emotions. Overall, participants selected the kind of interaction with controllers in which both hands and controllers were visualized as statistically the most preferred, and these effects are included in Chap. 5. For this research, work was divided between authors as Jan-Niklas Voigt-Antons was leading the idea from study design to paper writing, Tanja Kojić was responsible for conducting the study, writing-related work, and analyzing the results, Danish Ali developed the application. At the same time, Sebastian Möller supervised and contributed to the study design.

- T. Kojić, S. Ashipala, S. Möller and J.-N. Voigt-Antons, "Exploring Visualisations for Financial Statements in Virtual Reality", in 2020 IEEE International Conference on Artificial Intelligence and Virtual Reality (AIVR), IEEE, 2020, pp. 49–52 [24].

To find out how a 3D in VR visualization scenario affects an immersed user's ability to understand and interact with personal financial statement data, this paper compares the scenario where the user would be working with 2D personal financial statement data on paper to the scenario where the user would be working with 3D personal financial statement data in VR. This paper is based on data gathered for Sandra Ashipala's Master thesis, in which she conducted user studies and performed a basic analysis of results, while Tanja Kojić was responsible for further data analysis during the paper writing process and the study design was a collaborative effort by all authors. Results from this paper are included in Chap. 7.

- M. Vergari, T. Kojić, F. Vona, F. Garzotto, S. Möller, and J.-N. Voigt-Antons, "Influence of Interactivity and Social Environments on User Experience and Social Acceptability in Virtual Reality", in 2021 IEEE Virtual Reality and 3D User Interfaces (VR), IEEE, 2021, pp. 695–704 [25].

The impact of social environments and interactivity on user experience and social acceptability is investigated in this paper. Four social environments were created using 360° videos, and two virtual reality games with two degrees of involvement were created. The findings suggest that social environments and the degree of engagement should be considered while creating VR apps. This research is based on findings from Maurizio Vergari's Master's thesis, which were the outcome of collaboration between TU Berlin and Politecnico di Milano. Tanja Kojić was responsible for assisting with the user research and analysis of the results for the paper, while all the authors collaborated on the study design. As a result, the effects of the results are also incorporated in Chap. 6 of this work.

Furthermore, the author of the thesis Tanja Kojić was involved in activities at the ITU-T SG12, especially for the work items G.QoE-VR (cf. Chap. 2) and P.IntVR (cf. Chap. 2). These activities led to the start of the new and to the update of the corresponding Recommendation ITU-T Rec. G.1035.

ITU-T Contributions related to the work item G.QoE-VR (ITU-T Rec. G.1035):

- T. Kojić, J.-N. Voigt-Antons, and S. Möller, 11Contribution to ITU-T Rec G.1035: Contribution for Human Influencing Factors—Static and Dynamic Human Factors," ITU-T Study Group 12, Virtual-Geneva, ITU-T Contribution C.562, 2021 [26].
- T. Kojić, J.-N. Voigt-Antons, and S. Möller, "Update of Human IF for ITU-T Rec G.1035," ITU-T Study Group 12, Virtual-Geneva, ITU-T Contribution C.605, 2021 [27].

ITU-T Contributions related to the work item P.IntVR:

- T. Kojić, J.-N. Voigt-Antons, and S. Möller, "Proposal for new Work Item P.IntVR: Subjective Test Methods for Interactive Virtual Reality Applications," ITU-T Study Group 12, Virtual-Geneva, ITU-T Contribution C.548, 2021 [28].
- T. Kojić, J.-N. Voigt-Antons, and S. Möller, "P.IntVR: "Subjective Test Method for Interactive Virtual Reality Applications," ITU-T Study Group 12, Virtual-Geneva, ITU-T Contribution C.601, 2021 [29].
- T. Kojić, J.-N. Voigt-Antons, and S. Möller, "ITU-T P.IntVR: "Subjective Test Method for Interactive Virtual Reality Applications"—Quality of Experience aspects for Interactive Virtual Reality," ITU-T Study Group 12, Geneva, ITU-T Contribution C.060, 2022 [30].

1.4 Thesis Structure

As an introduction, Chap. 2 gives an overview of the current state of the art concerning VR research, serious gaming, and UX. This chapter also gives a brief overview of evaluating the most important parts of the UX in VR and serious gaming, as well as the current standardization of influencing factors for VR. Based on in, in this chapter, the final RQs are formed and presented for this work.

In Chap. 3, an overview of the user testing methodology is presented, as well as a description of the set-up and the VR applications. Each of the five VR applications is discussed here, along with the required hardware and software. Furthermore, the processes for each of the eight user studies based on one of the linked RQs are explained, as are the study conditions, the parameters evaluated using the questionnaires, and a description of the participants who took part in this research. Finally, for each user study, this chapter summarizes all the main attributes in the form of an overview table.

The user research findings are presented from Chaps. 4 to 7, following the structure established by categorizing influencing elements for VR services. At the end of each chapter, the results, discussion, and relationship of UX components are provided, along with proposed design guidelines. Chapter 4 presents results from Studies 1, 3, and 4, with an emphasis on influences depending on the VR device itself and network delay. Significant differences in several components of the UX model, such as different feelings of presence, flow, perception of delay, readability, and support level, as well as significantly different levels of overall usability and quality of experience, are reported in the results. Although the content displayed in a VR system can be considered part of the System Influencing Factors, it has been an important part of the results, so it is discussed separately in Chap. 5. Interactions and environmental design are two subfactors that are examined. This chapter reports on key influences on components of the UX model and is based on Studies 5 and 6. In this part, an additional investigation is described regarding text as a design element, reporting on interesting results additional to UX, concerning limitations of current standards for readability. In terms of context, it was limited by the methods used in laboratory experiments. Therefore only one subfactor was explored and given in Chap. 6, which was the social subfactor investigated in Studies 1 and 8. Despite the limitations, the data presented in this chapter demonstrate interesting implications on UX components such as a feeling of presence and social presence, emotional responses, and social acceptability. Next, Chap. 7 focuses on human factors, demonstrating how users of different ages, genders, and previous VR experiences perceive and rate UX for VR serious gaming. These influences are explored in Studies 2 and 7, and the results concerning human variables are reported.

The final Chap. 8 summarizes all results and presents an overview of answers based on each RQ's results. In addition, the limitations of this work are summarized in this chapter, along with future possible improvements of UX for VR serious gaming.

Chapter 2
Related Work

2.1 Virtual Reality and Serious Games

The so-called second wave of VR has brought to research and market a lot of new displays, input devices, and content solutions during the last few years. Not only has new hardware entered the consumer market with low-cost price patterns, but whole new technologies are also being designed and developed.

Sutherland's concept for the Ultimate Display [31], which would reproduce the actual world in every sense, has been around for more than 50 years. Similarly, it has been a long time since Jaron Lanier came up with the term *Virtual Reality* [32] to try to bring all of the different ideas about sensory stimulation together. This produced a lot of enthusiasm, but it also brought the scientific community together to work on technology and algorithms to make Sutherland's vision a reality. Even while public attention faded after a few years, researchers saw the potential and continued to work on it. Stereoscopy, a field of vision, and synchronized multimodality appear to be vital components of the VR experience since they make individuals feel like they are present and immersed [33], which are referred to as two of the most significant aspects of the experience. Aside from a broad angle field of view and three-dimensional vision, features that are likely to help people feel more immersed and present may also incorporate audio or tactile inputs. Therefore, over the years, VR has been characterized by the public as a medium usually connected to a specific set of technological devices, such as computers, head-mounted displays, headphones, and motion-sensing gloves [34]. However, the focus was broadened over the years to include not only technical aspects but also experimental aspects as well as human perception of such systems.

T. Kojić, *User Experience for Serious Games in Virtual Reality*, T-Labs Series in Telecommunication Services, https://doi.org/10.1007/978-3-031-75530-9_2

2.1.1 Definition of Virtual Reality

Variety and heterogeneity are seen in the range of VR definitions, and the literature has a wide range of relatively variable definitions of VR developed over time and evolved as a result of technical improvements [35]. Therefore the term *Virtual Reality* can be used with many different meanings, from those stating that imagination can be considered as a form of VR [36], over defining it as any experience in which the user is effectively immersed in a responsive virtual world [37], to those defining it based on technology used to simulate reality.

The Reality-Virtuality continuum was introduced by Milgram et al., and it illustrates the position of the technology on a continuous scale ranging from entirely real to completely virtual [38] as shown in Fig. 2.1. Based on it, the most common terms are Virtual Reality (VR), Augmented Reality (AR), Augmented Virtuality (AV), and Mixed Reality (MR). All four of them use different degrees of immersion by simulating part of reality.

Some of the recent definitions for the terms are the following:

- *Augmented Reality*: "AR overlays computer generated content on real space" [16].
- *AR technology*: "A technology that superimposes a computer-generated image on a user's view of the real world, thus providing a composite view" [40].
- *Augmented Virtuality*: "AV incorporates real objects into a virtual space" [16].
- *AV technology*: "A technology that modifies a Virtual Environment by augmenting it with real data" [38].
- *Mixed Reality*: "MR combines real and virtual content, registered in various virtual spaces, including 2 dimensional (2D) and 4 dimensional (4D), allowing for real-time interaction; according to perspective" [16].
- *MR technology*: "A blend of physical and virtual worlds that includes both real and computer-generated objects. The two worlds are 'mixed' together to create a realistic environment. A user can navigate this environment and interact with both real and virtual objects. Mixed reality combines aspects of virtual reality and augmented reality. It is sometimes called 'enhanced' AR since it is similar to AR technology but provides more physical interaction" [41].
- *Virtual Reality*: "VR occludes physical space to provide interactive and non-interactive experiences of a fully computer-simulated virtual world or a photographically captured real world." [41].

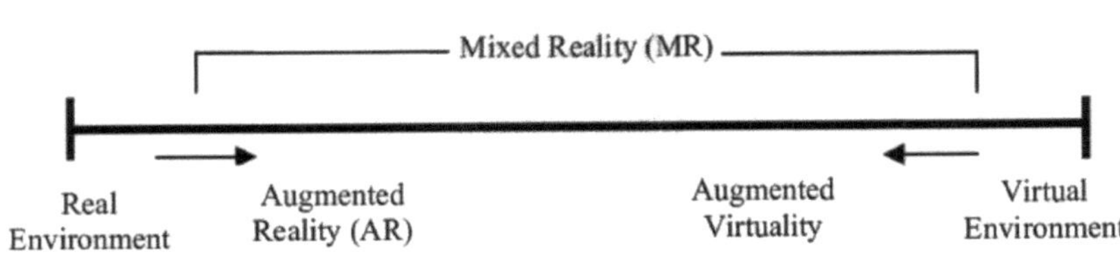

Fig. 2.1 Reality-virtuality continuum [39]

- *VR technology*: "The computer-generated simulation of a three-dimensional image or environment that can be interacted with in a seemingly real or physical way by a person using special electronic equipment, such as a helmet with a screen inside or gloves fitted with sensors" [42].

It should be noted that some of the above definitions are limited to the visual medium, while others are not. Similarly, some definitions are unaffected by hardware type, but others are more specific. Furthermore, other definitions include the amount of immersion, with the main objective for VR being to provide a higher level of immersion, whereas AR focuses on experiences embedded in an actual environment. The definition of immersion and sense of presence perceived will be further explained in Sect. 2.2.3.

Lastly, even though the standardization of definition in research has not yet been finished, the ITU defines VR as:

- *"A technology that uses game engines (e.g., Unity) to create artificial environments that enable people to interact in six degrees of freedom (DoF). VR services aim to provide users with high levels of immersion and presence wherein users may feel detached from their physical, real-world surroundings. This is different from augmented reality (AR) or mixed reality (MR), which enhance user experiences by adding virtual components such as digital images, graphics, or sensations as a new layer of interaction with the real world*" [15].

Within this work, for VR definition, it will be referred to as the one given by ITU for VR technology [15], leaving AR, AV, and MR to future investigations.

2.1.2 VR System

2.1.2.1 Market Solutions and Trends

Many VR devices have been released on the consumer market in recent years, allowing VR technology to gain popularity. The worldwide VR industry is expected to grow from less than five billion US dollars in 2021 to more than 12 billion US dollars by 2024 [43]. The predicted growth will likely benefit both the business and consumer segments, including the growing development of the VR gaming industry. Despite the fact that research and hardware for VR have been available for a while, a new wave of market-ready VR solutions began only a few years ago. This is consistent with the fact that the products that have begun to arrive on the market are now both useful and affordable.

When it comes to the types of VR devices on the market, it is reported that standalone HMDs take 49.8% of all VR headsets delivered worldwide in the first quarter of 2020 [43]. In terms of popularity, the Oculus VR headsets, particularly the Oculus Quest 2, are the most popular VR headsets now available on the market [43]. Additionally, other advancements to VR technology, such as making devices smaller and more fashionable, will probably as well influence the expanding number of VR headsets that are currently in use.

With different VR devices on the consumer market, it is expected that VR technology will be used in many different fields. The gaming and entertainment business is the most prominent, but retail and e-commerce, education and training, travel and tourism, advertising, healthcare, automotive, architecture, design, and many more industries are also embracing VR solutions [44].

2.1.2.2 Hardware

To display virtual scenes, VR technology uses a variety of technical approaches and configurations [45]. The Cave Automatic Virtual Environment (CAVE) and Head Mounted Display (HMD) are two of the most prominent VR technology solutions [46]. The biggest difference between the two is how the virtual environment is represented, with different levels of isolation for users and hence a different feeling of immersion. The differences between the two are explained with the following descriptions:

Cave Automatic Virtual Environment is an immersive VR environment [47] in which stereoscopic images are projected onto three to six of the walls of a room-sized cube [48]. It is also usual to wear 3D glasses with shutters that open and close in line with the visuals to create the illusion of depth and use at least one controller or remote to interact with the environment.

Head-Mounted Display is a head-mounted display device with a tiny display optic in front of one (monocular) or both eyes (binocular) [49]. This display is designed to completely immerse the user in whatever experience the display is intended for since it assures that no matter where the user's head turns, the display is directly in front of the user's eyes [50].

Over the years, HMDs have experienced hardware issues that have caused individuals to feel ill, such as motion sickness or eyestrain [51]. As a result, the majority of previous research concluded that there was no difference between CAVE and HMD or that CAVE was a better go-to solution [52]. However, current hardware solutions and studies indicate that this tendency is shifting. HMD technology and how people use it have improved in recent years, and HMD solutions are currently advancing considerably faster compared to the CAVE technology [53]. Therefore, in this chapter, the focus was put on the HMD solutions.

2.1.2.3 Input and Output

Multimodal interactions are essential to the VR experience and can involve some or all of the human senses from sight and hearing to touch, smell, and even taste [54]. Which and how many interactions will be included in the VR experience depends on the idea of the application itself and on the computer platform used. In order to include more interactions, it is necessary to have special devices that can simulate the virtual world in a timely manner in relation to human input. This is why it is essential to have systems that can process information quickly and are specifically developed to handle input-output synchronization [55]. Input-output device technology ranges from HMD devices to various versions of haptic interfaces, which integrate force and tactile sensations [56]. Virtual reality input-output devices cannot work alone and must be used in conjunction in a single simulation system.

Input devices are used to interact with VR to capture information about the user such as movement and position, to desired interactions in the virtual environment. The data used by the computer system are collected from input devices. This information can be as minimal as just moving eyes, head, clicking with hand or up to full physical movement. Such devices are most often controllers that come with HMDs or joysticks, but may include various track pads, gloves or belts with sensors, devices for running, rowing or cycling, and even overall body suites completely covered with sensors [54].

On the other hand, output devices are any piece of hardware that allows data to be sent from the user's computer. Such devices receive processed data from the computer and then convert it into the form of audio, video, or physical copies that the user can understand. Such devices for VR are usually monitors and projectors for video, and headphones and speakers for audio. In addition, by using mechanical or electrical stimuli, such as vibrations or application of force, haptic output devices enable the user to feel physical interaction with virtual things, and haptic devices are most often implemented together with input devices such as gloves or mechanical sensors [56].

Current market systems often include input-output components [54], such as *position tracking*, which typically begins with the detection of the user's location, such as the position of the user's head and hands. Furthermore, *sensors*, in addition to trackers, are used to input instructions into the virtual world. All user inputs are combined with geometric, physical, and behavioral models of various virtual objects. Depending on the system configuration, the user is then given *visual feedback*. The user will see feedback on the screen (i.e., HMD or CAVE), and most VR systems will enhance visual input with *audio feedback*, but feedback may be only audio in certain cases. Additionally, *haptic feedback*, which employs pressures to create feedback during an engagement, can also be used. The interaction loop between input and output is immediate, and devices are frequently employed for both input and output.

2.1.2.4 User Interface

A system element that facilitates the communication between the user and the computer is *User Interface (UI)*. This communication may occur through hardware like HMDs and displays, as well as specialized software with graphics and menus [57]. This is made possible by the interface, which first turns the user's actions into an interpretation that the computer can comprehend and then translates the

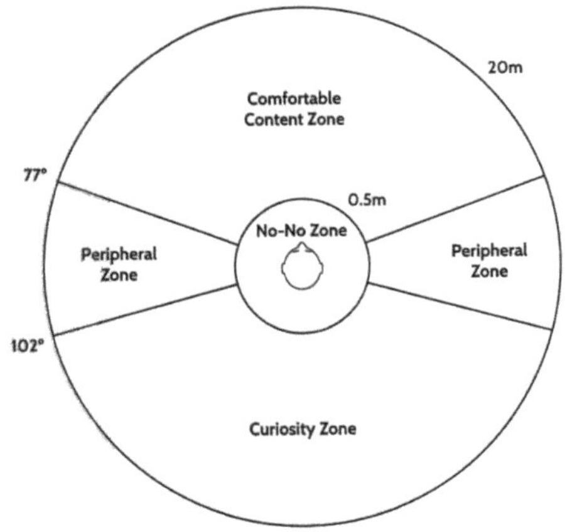

Fig. 2.2 UI viewing zones in VR [61]

computer's actions into an interpretation that the user can understand and act on [58, 59]. The advancement of graphics and virtual environments has had a significant impact on the rise of three-dimensional user interfaces [60].

In VR, user interfaces are not constrained by screen size. Instead, they may be included in the virtual environment, which might have an influence on how the user feels. The view in VR is separated into three distinct zones, which are defined by the position of various components around the user [61]: The comfortable *content zone* is a 140°-wide angle in front of the user; the *peripheral zone* is between 70° and 105° angle to the user; and *curiosity zone* is anything else behind the user, as shown in Fig. 2.2.

The distance between the UI and the user is also important and is defined to be at least 0.5 meters, and at most 20 meters since any further than that will appear to be beyond the horizon. Also, the vertical placement of the UI is explored, and it is determined to be approximately a 30° angle from the user's straight viewpoint, which is considered as 0°. This is shown in Fig. 2.3 done so that the vertical positioning of the UI may be optimized.

There are more than several examples where VR user interfaces have risen in popularity because they can be used in locations where regular desktop computers cannot [62]. In particular, such examples are three-dimensional scientific images, where users with VR interfaces may engage in a more immersive manner [63]. It is difficult to see a lot of information on a typical screen without missing important details or chunks of it. Data visualization makes data analysis easier by visually displaying information to users [64]. Furthermore, data visualization has been used in the field of personal finance to construct relevant dashboards that allow users to keep track of things like their spending patterns and income. In contrast to more conventional data visualization methods, VR offers an additional dimension to the process, which may make it more helpful to the user in enabling them to comprehend better the data they are seeing [65]. VR has the potential to transform what would normally be a monotonous banking transaction into one that is exciting and interesting.

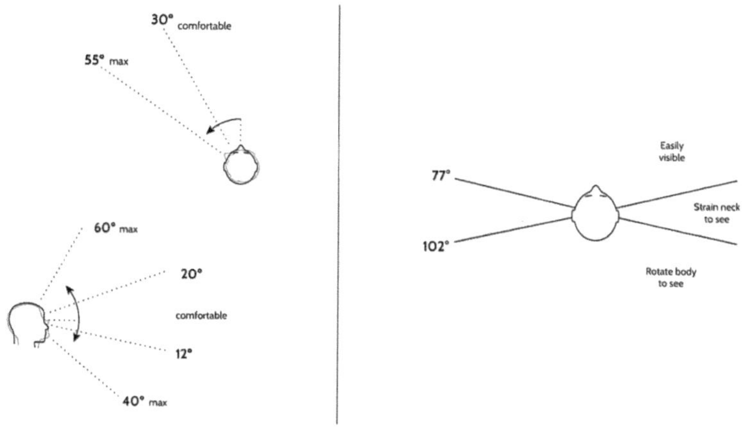

Fig. 2.3 Comfortable head movements in VR [61]

Design guidelines like those described in Figs. 2.2 and 2.3 are made for VR interfaces in general, but when it comes to UI research in VR exergaming, the latter is restricted by the lack of VR exergames. One of the first VR exercise games was PaperDude—a VR bicycle-riding exercise game. The player must ride a bicycle through the neighborhood while avoiding obstacles while delivering newspapers. Cycling is the most popular type of sport in VR exergames because people already know how to do it and are less likely to get motion sickness. However, research is not limited only to one type of it. When making a UI for a VR exercise game, the focus should be on how easy it is to use and how the user feels while working out. Since this is such an important factor, many studies have been done on different sports, such as biking [66], running [67], or rowing. However, those research studies had not yet taken a thorough look at the components that influence visualization or included biofeedback in the scenario.

When it comes to the research that focuses on player view in VR, two possibilities are defined [68]: third-person or first-person viewpoint. Third-person view, often known as virtual body UI, refers to a player seeing a virtual version of themselves as an avatar[69]. The first-person viewpoint implies that the player only sees the view, not his body. Because the vision in VR is matched to the location of the player's head, the first-person viewpoint is a logical option for VR. However, the rest of this research is primarily concerned with the issues related to the use of immersive technology in exergaming, such as motion sickness, motion tracking, and physical feedback with health or safety consequences, rather than the UI itself.

2.1.2.5 Interactivity

Interaction with and in VR is a broad field that can include everything from general motor learning and control concepts to remapping concepts and discussions of social interaction mechanisms that allow users to interact directly with one another. Interactivity is important in allowing users to feel truly free to participate in the virtual environment. *Interactivity* in VR "is not merely the ability to navigate the virtual world; it is the power of the user to modify this environment. Moving the sensors and enjoying the freedom of movement do not themselves ensure an interactive relation between a user and an environment: the user could derive his entire satisfaction from the exploration of the surrounding domain. He would be actively involved in the virtual world, but his actions would bear no lasting consequences" [70]. Previously, technology was not as advanced, resulting in less satisfactory user experiences. New technological advances, research, and development have now elevated users' sensory outputs to an entirely new level.

Overall, three major factors influence VR engagement: speed, range, and mapping [34]. The speed of interaction, also known as response time, relates to how quickly information may be absorbed into a virtual world, and it is one of the most important characteristics of an interactive system. User activities should be visible in real time and have the ability to alter the environment in an instant. Such real-time interactivity is the level that new media is attempting to adopt, allowing for a well-timed change in perspective of the virtual environment. The range determines how many different interactions are offered to the user at any given time. Accuracy, or how precisely and correctly the virtual environment replicates the real world, is another necessary aspect of the range. The final component is mapping, which refers to the system's capacity to map its controls to changes in the virtual world naturally and predictably. There are two mapping functions: one for interaction between the actual and virtual worlds and another for mapping how the actions of those controls correspond to activities within that environment.

However, the scope of this work will be limited to the concept of interactivity, specifically interaction with hand tracking and biofeedback.

Haptics and Hand Tracking

One of the benefits of virtual experiences is an immersion into other worlds [71], but any touch with the actual world, via sensors or controls, might shatter this illusion [72]. As a result, new HMDs include capabilities that let you interact with them using your hands rather than controls. This new option might lead to some intriguing applications of VR in gaming and other places where serious games are applied. VR is already being used to learn in a variety of settings [73], including in medical [74] and manufacturing [75]. Precision has always been vital in these disciplines, as have hand motions. But it is also important to ensure the user has a positive experience to obtain quality simulations.

Realistically showing the user's hand in the virtual world is an important part of the VR experience [76]. This is not just about VR; it has also been studied in gaming with other types of virtual environments [77], with a focus on algorithms for how to get accurate hand tracking through different sensors [78]. In a study about hand tracking and visualization in a VR simulation, a design was made so that the fingers and palm could be tracked [79]. The study focused on experiments where people had to copy certain hand positions and then report on how fast they did it. Despite this, there has not been a lot of work done to improve the user experience of hand tracking in VR, as this has started to gain the attention of researchers in recent years. Presumably, no research has been carried out with built-in hand tracking in HMD that focuses on user experience with various types of interaction.

Biofeedback

Measurement and monitoring of physiological processes that are rendered perceivable for human senses in order to assist individuals in altering the outcome of their biological processes are included in the definition of the term *biofeedback* [80]. Its primary application is either in the context of physical rehabilitation or in the context of sporting activities, during which it is intended to improve both the heart rate and the respiratory rate [81]. However, there have been several studies that have previously been conducted on the topic of combining VR with learning through the application of biofeedback. In a study, researchers have created a biofeedback training program consisting of 10 sessions that were carried out with great success on a golfer who participated in a study. In this example, researchers successfully incorporated biofeedback on heart rate variability into a VR environment. While the majority of physiological data study uses it as a measurement tool, some researchers have investigated its influence when direct and voluntary physiological control is employed as a way of engagement [82]. Moreover, it was shown that the direct biofeedback method is quite engaging. Also, the richness of biofeedback as a direct game component inspires new ideas for customized immersive experiences.

2.1.3 Serious Games

The use cases for VR have extended into more sectors, and VR technology can be used to make games that are not just fun and entertainment, meaning it can be used to create serious games:

- *Serious games* "are games that do not have entertainment, enjoyment, or fun as their primary purpose; and are created with the intention to achieve at least one additional goal" [83, 84].

It should be noted that serious games are not a specific form of game. Serious games include different forms and scenarios, so they can be designed as action-adventure games, strategy games, sports games, etc. Also, serious games are not the same as gamification, as gamification is the incorporation of game approaches or aspects into nongame applications and processes [85].

Previously, research has demonstrated that serious games may be used in a variety of contexts: from healthcare, where serious games are used to develop and train new technical abilities, to education, where students do schoolwork better through various interactions compared to the traditional non-immersive approach [86]. Serious games are activities that are intended to be entertaining while also teaching and training users in certain areas and activities. Serious games thus take a different approach to education than typical classrooms, where the teacher controls the learning since, in a serious game, the trainee feels in control of an interactive learning process, which increases the likelihood of active and critical learning [87]. As a result, serious games are commonly implemented in teaching and training [88].

The objectives of serious games can be used to categorize them. Exergames, for example, focus on making users physically active [89], whereas advergames are used for marketing or recruitment and may increase players' awareness of specific themes [90]. Today's serious games also aim to help people make medical diagnoses, understand how causes work, run businesses, make decisions, improve their social skills, learn how to solve conflicts, change their lifestyle habits, and do a variety of other things [84].

Finally, VR serious games share the goal of serious games by using the VR technology and should improve users' experiences and help them learn more [88], but immersive VR environments bring new challenges regarding how to create serious games that operate effectively in these situations.

2.1.3.1 Exergaming

Exergaming, a term derived from the words "exercise" and "gaming," was defined by Oh and Yang as:

- *Exergaming* "is the activity of playing video games that require exertion or movements that are more than sedentary activities and also include strength, balance and flexibility activities [89]."

Exergames, like any other traditional game, may be configured to be single or multiplayer, with game synchronization over an Internet network. They can either be based on video games or designed specifically for VR technology. Also, exergames can be used for a variety of purposes, ranging from sports-related to rehabilitation-focused. There have already been various sport-based applications on the market, ranging from those aimed at beginners to those aimed at pros and their competition preparation.

Some examples of research exploring user experience and performance with sporting applications include, but are not limited to, table tennis and rowing. Table Tennis for Three is an exergame in which each participant has the same table tennis set-up as the other [91]. A ball, a paddle, and a customized table with a backboard onto which the virtual environment is projected to form the set-up. Another study focuses on rowing as an exergame, in which an ergometer is combined with VR [92]. The focus of this investigation was the impact of a companion portrayed as an avatar on performance, and it was discovered that VR increases performance for it. Furthermore, when it comes to the use of exergaming in rehabilitation, it

has been proven in several different programs, addressing all ages of users, from kids to the elderly. Although children are generally more open to the concept of exergaming, which makes it an especially useful tool for encouraging kids to concentrate on getting better [93], the concept of exergaming may benefit people of all ages. The need for remote rehabilitation options has been highlighted, especially during the COVID-19 pandemic. Exercise games have also been demonstrated to be a beneficial supplement to rehabilitation programs, such as those that assist the elderly in their regular treatments [94].

VR technology might be beneficial for exergaming in a variety of ways, and the design options are wide. However, along with design possibilities comes design responsibility to make such environments regulated and tailored for enhancing user performance and motivation while maintaining context safety when conducting workouts with an HMD set-up. With that objective in mind, it is necessary to understand what factors influence the experience of serious games in VR.

2.2 User Experience Research

For a long time, the primary focus of Human-Computer Interaction (HCI) research was on increasing system effectiveness, and efficiency [95]. Nowadays, the goal is not only to avoid a computer crash [96] but also to enable positive feelings when using an interactive system and even to design systems that contribute to a particular emotional experience [97]. In addition, HCI research has shifted away from the conventional paradigm of cognitive information processing [98] and toward notions like emotional design and affective computing, [99]. The phrase "User Experience," a term popularized by Donald Norman's work at Apple Computers [100], became widespread.

Because of the great effort and work on the definitions and frameworks presented in the research on UX, there is not just one common definition for it, as the research on UX and related themes is extensive and diverse. However, the International Organization for Standardization (ISO) introduced the revised ISO 9241–210 standard in 2010:

- *User Experience* is defined as "a person's perceptions and reactions resulting from the usage or expected use of a product, system, or service."

Even though this definition leaves rather some space for interpretation, it can be used in many different cases and was used later on for defining particular factors of UX. Meaning that UX depending on the context is used to describe a variety of different things, including subjective, complex, and dynamic scenarios. This is because the term UX is used to describe the result of all three factors [96, 101]:

- System properties
- The context around the user and system
- The user's state

Furthermore, while addressing "users" and their "experience" with technology, it is important to recognize Quality of Experience (QoE), comparable to the field of UX, but frequently used inaccurately as a synonym for UX. Even though concepts come from different backgrounds (UX from usability and HCI, while QoE has its roots in telecommunication) [102], during the previous decade, QoE and UX have been essential ideas for the design and assessment of products, systems, and services [103], although there are some similarities and contrast between the two:

- *Quality of Experience* is defined as "the degree of the delight of the user of a service. In the context of communication services, it is influenced by content, network, device, application, user expectations, and goals, and context of use [104]."

In the UX community, the idea of *experience* and what it means has been carefully and critically analyzed. As a result, evaluation of experiences and characteristics of the experience have become central parts of UX practice [102]. In the field of QoE, on the other hand, the focus is more directly on how quality is made and on the things that make quality seems good (called quality features [105]). That is why the focus is on judging quality and less on judging experiences and the parts of experiences that affect their quality [102]. The emphasis on human centering is one of the most important and general aspects of UX. Technology does not drive UX [101], which is true both in theory and in practice. QoE, on the other hand, is still sometimes dependent on service quality [104], and most QoE research is focused on systems and technologies. In contrast to QoE, which focuses on evaluating the experienced quality of a system, measuring the experiential characteristics in UX involves considering a number of additional factors. UX evaluates aspects such as affect, emotion, fun, aesthetics, hedonics, and flow [106]. As a result, this method of evaluation often considers a considerably wider variety of measurable characteristics or experience dimensions.

The User Experience in the User Experience in Immersive Virtual Environment (UXIVE) model [107] shows the main components of UX in VR and their interactions with one another and includes ten main parts:

- *Presence* is defined as the user's "sense of being there" in the virtual environment [108]. It is further classified as telepresence, self-presence, and social presence in a virtual environment [109].
- *Immersion* is defined as the "objective degree of sensory fidelity provided by a virtual reality system." "Complex technologies that substitute real-world sensory information with synthetic stimuli" create the immersive aspect in a virtual environment [110].
- *Engagement* is defined over-involvement as "a psychological state experienced because of focusing one's energy and attention on a coherent set of stimuli or meaningfully related activities and events" [111].
- *Flow* is defined as "the holistic sensation that people feel when they act with total involvement" [112].

- *Usability* is defined as "the ease of using (i.e., efficiency, effectiveness, and satisfaction) the virtual environment." The word "ease of use" refers to how difficult it is to use a certain system [107].
- *Skill* is defined as "the knowledge the user gains in mastering his activity in the virtual environment" [107].
- *Emotion* is defined as the subjective feelings (i.e., joy, pleasure, satisfaction, frustration, disappointment, and anxiety), and it refers to how a person feels when navigating a virtual world [113].
- *Experience consequence* is defined as the symptoms the user can experience in the virtual environment, and it includes symptoms such as simulator sickness, nausea, headaches, and eyestrain [114].
- *Judgment* is defined as "the overall judgement (i.e., positive, indifferent or negative) of the virtual environment" [107].
- *Technology adoption* is defined "as the actions and decisions taken by the user for future use or intention to use of the virtual environment" [107].

The first three features are especially important for this work and will be further explained within this chapter.

In the context of this thesis, certain aspects of these technology-driven and technology-centered QoE techniques will be blended with existing UX approaches. However, the latter will be the primary focus of the research presented because it focuses on the effects of parameters on users from a human-centric perspective, providing a source for the improvement of design for VR serious gaming.

2.2.1 Evaluation of User Experience

Choosing the most appropriate UX questionnaire for a certain assessment project is not always straightforward, as there is a wide variety of accessible UX questionnaires [115]. Each of them measures a subset of some concept of UX based on its scales and items. There are several existing surveys, each designed to investigate a subset of UX features. Some are suitable for a wide variety of product categories, whereas others are highly specialized for certain product types or characteristics. Many traditional HCI techniques and approaches, such as Goals, Operators, Methods, and Selection Rules (GOMS) [116], Usefulness, Satisfaction, and Ease of use (USE) [117], or surveys, such as the well-known System Usability Scale (SUS) [118], excluded aspects such as hedonic aspects or emotions.

The System Usability Scale (SUS) is a straightforward and dependable technique for assessing usability. This is a ten-item survey with five answer choices ranging from strongly agree to strongly disagree [119]. It supports the evaluation of a wide range of goods and services, including hardware, software, mobile devices, websites, and apps, including those for VR. Over 1300 papers and publications reference SUS, which has become an industry-standard [120]. The advantages of using SUS include the fact that it is a simple scale for participants to administer, that small sample sizes may be used effectively, and that it can successfully distinguish between usable and unusable systems.

Usefulness, Satisfaction, and Ease (USE) are the three aspects that appeared most prominently during the early stages of USE Questionnaire development. USE is a 30-item questionnaire that analyzes four elements of usability: utility, ease of use, ease of learning, and satisfaction [117]. Because it is publicly available and technology neutral, this metric is suitable for a wide range of usability assessment settings, including VR. Items in it have descriptions that are clear and make sense, but not much research is done with it, as it is not so popular compared to the other tools.

Furthermore, approaches from a variety of fields were investigated, modified, and implemented—for instance, from psychology, the Self-Assessment-Manikin (SAM) [121] and the Repertory Grid Technique [122] were incorporated. In addition, research reveals that, aside from generic UX, emotions and effect are the most examined aspects in UX research. In this context, the SAM was shown to be the most often used instrument, confirmed as well by another evaluation [106] that found that the AttrakDiff [123] and the SAM were particular favorites. In addition, this study found that there is a shift in UX research from quantitative to qualitative research, with many studies employing both methodologies. AttrakDiff is a questionnaire focusing on how people subjectively evaluate an interactive product's usefulness, and aesthetic attractiveness [123]. The assessment data are used to determine how appealing the product is in terms of usability and aesthetic appeal, as well as whether or not it needs to be optimized. AttrakDiff evaluations distinguish between pragmatic and hedonic quality, having the resulting attraction as the product of pragmatic and hedonic variables. Likewise, SAM is a brief self-reporting questionnaire that measures enjoyment, arousal, and dominance. Due to its non-verbal character, the questionnaire is accessible regardless of the respondent's age, language proficiency, or level of education, making it popular in UX research.

Another often used questionnaire for analyzing users' subjective opinions regarding product UX is the User Experience Questionnaire (UEQ) [124]. The UEQ's objective is to allow end users to do a quick assessment that provides a comprehensive impression of UX. It should allow consumers to communicate their thoughts, opinions, and attitudes about the studied product in a simple and quick way. The UEQ consists of 26 questions, with a shortened version available. The UEQ-S [125] is a small eight-item variant of the UEQ that is optimized for usage in circumstances when the full UEQ cannot be applied, as the UEQ-S misses to analyze the UX qualities such as Attractiveness, Efficiency, Transparency, Dependability, Stimulation, and Novelty, which are covered in the UEQ.

A literature review compared the academic uses of the standard questionnaires AttrakDiff and UEQ [126]. The results show that the use of standardized questionnaires has been steadily growing since 2006 when the first articles about their use were published. AttrakDiff is still very popular; however, in terms of usability, the UEQ questionnaire has recently outperformed AttrakDiff. Nonetheless, both are arguably the most commonly used UX assessment methods in the field.

2.2.2 *Motivation and Engagement*

When it comes to VR, motivation and engagement are frequently mentioned terms, particularly when it comes to motivation in gaming [127] or learning in VR environments [128]. Overall, psychological research has discovered a link between motivation to perform a task, such as learning, and success at it [129]. So, the idea behind using VR is to enhance performance by making people more self-motivated because, in VR, they can enjoy experiences that they cannot have in real life, and all of this by giving them the ability to control and think in virtual environments [130]. When people devote more time, work, or learning effort, they remember it better and for a longer period of time [131]. Therefore, motivation is important for engagement, but engagement in virtual environments is also determined by how people feel while exploring VR worlds.

The Self-Determination Theory (SDT) [132] is a psychological idea that describes how individuals are in command of their own lives and how they select what to do. This theory explains not just why individuals make the choices that they do but also how they feel about the outcomes of those actions. According to this theory, the majority of individuals are motivated by the need to improve themselves and find happiness in their lives. The SDT framework is predicated on the three fundamental psychological requirements that are shared by all people: competence, connection, and autonomy. It is assumed that people will be able to become self-determined if all three of these conditions are fulfilled. This theory and other SDT principles have been used to explain amotivation and motivation in a variety of contexts, such as education, work, health, parenting, and athletics, among others. Video games have also made use of this principle in various ways. The outcomes of a study suggest that SDT's views on the criteria for autonomy, competence, and relatedness may be able to predict how much a person would enjoy a game in the future and how much they will play it [133]. Based on the SDT, there are five types of motivation:

- *External Regulation* is referring to the impact on motivation that comes from outside of the individual. A person is said to be motivated from the outside when they are acting in some manner in order to achieve the reward that is offered from the outside.
- *Introjected Regulation* refers to behavior caused by emotions such as shame or guilt; in this situation, a person acts or refrains from doing it out of fear of obligation, which is tied to their own perceptions of rewards and punishments.
- *Identified Regulation* is referred to when the importance of behavior is viewed as helping in the achievement of a personal goal, but not because the individual enjoys performing it.
- *Integrated Regulation* is a type of extrinsic motivation in which the person's own values and needs are taken into account when deciding on strategies.
- *Intrinsic Motivation* occurs when a person is motivated only by his or her own personal interest, satisfaction, and pleasure. There is no need for fear, a reward,

or any other external element. Intrinsically driven activities are ones that are done just for the enjoyment of the action and how it makes the person feel.

When attempting to explain why people make the decisions they do, these types of motivation are frequently invoked as possible explanations. For the purpose of explaining the factors that influence an individual's decision-making process, SDT is frequently applied in different research fields. For example, in sports, in order to determine what motivates users, questionnaires like the Sport Motivation Scale (SMS) [134] are typically used. This scale assesses people's motivation for participating in sports activities and comes in three versions: the original with 24 items, the new version with 18 items, and the short version with only six items. Another example is the Motives for Physical Activity Examine - Revised (MPAM) [135]. This questionnaire is designed to examine the importance of five main reasons for participating in physical activities. It can be used in a variety of sports, including weightlifting, aerobics, and team sports. Based on the SDT's theoretical foundation, the scale is made up of 30 items separated into five categories: fitness, appearance, competence and challenge, sociability, and enjoyment.

Furthermore, the flow theory is often associated with the idea of engagement [136]. Flow is attempting to strike a balance between boredom and fear, and as a dynamic experience of a person becoming one with his or her activity, it is a balance between boredom and anxiety [112, 137]. To assess flow, different questionnaires are usually used depending on the scenario, so there can be made particularly for gaming like GameFlow [138], while the Flow State Scale (FSS) is often used to assess flow in physical activities [139]. After all, gamification grew out of the concept that games could be used to get people interested in non-entertainment areas like education, health care, and sports.

2.2.3 Presence and Immersion

There are many different ways to understand what it means to be "present" in the virtual world or to be "immersed" in it. When describing experiences with VR, the terms presence and immersion are sometimes used interchangeably. This work does not undertake in-depth research on the term presence, and it thus considers the provided definition from the UXIVE model as a standard reference for the core concept [107]. However, the literature provides the most current and precise definitions [140]. Furthermore, in literature, there are defined three different types of presence [141]:

- *Telepresence* is defined as "the degree to which one feels present in the mediated environment as opposed to the immediate physical environment" [34].
- *Self-presence* is the degree to which the "virtual self" is seen as the "real self" [142].
- *Social presence*, also known as *copresence*, is defined as the "feeling of being with another person" [143].

Overall, telepresence is based on how strongly the user feels the environment and space of the virtual world. Self-presence, on the other hand, is different from telepresence because it is based on how connected users feel about their virtual body, emotions, or identity [144]. Finally, the concept of social presence was originally proposed as a means to describe how individuals communicated with one another via various forms of media [145]. Social presence is an essential component of online communities that bring people together. Without it, the mediated other is seen as a machine rather than a human [146].

When it comes to measuring presence as part of UX research in VR, there are several different approaches, from behavioral and physiological measures to the use of questionnaire [147]. However, questionnaires are the most often used method. Before completing a questionnaire on their experience, participants are requested to observe or engage in a virtual environment. The questions are answered using ordinal scores, often ranging from 1 (no presence) to 7 (a strong feeling of presence)[147]. Some questionnaires that are based on this metrics are the Presence Questionnaire (PQ) [148] and the Igroup Presence Questionnaire (IPQ) [149], as all use a Likert scale with a 1–7-point scale. The PQ has 29 items divided into four components Involvement, Sensory Fidelity, Adaptation/Immersion, and Interface Quality, while the IPQ consists of 14 items with the subscales General Presence, Spatial Presence, Involvement, and Realism. Depending on what aspect is the focus of the study, one or the other is often chosen within VR research.

Questionnaires have several advantages over other techniques of measuring presence. Overall, they are affordable, easy to use, and applicable for work with VR. Further, there is no need for additional scientific equipment, as is usually the case with physiological measures. However, the absence of a clear definition for presence as a psychological term may make questionnaires difficult to use. In particular, comparing the study results using different questionnaires may be challenging [150]. Lastly, when asked explicitly or indirectly about presence in a questionnaire, respondents may provide responses they would not have given otherwise, which makes presence-measuring questionnaires sensitive to response bias [151].

As the typical presence questionnaires are not measuring the social presence or have only as a dimension, there have been as well specialized questionnaires developed for measuring social presence in virtual environments [152]. One of the most commonly used measures is Networked Minds Questionnaire [153] attempting to assess how much attention the user believes their partner or participant is paying attention to them, how their emotions influence them, and how well they understand each other.

2.2.4 Simulator Sickness

VR sickness, also known as a simulator, simulation, or cybersickness, occurs when a person is exposed to a virtual environment and experiences symptoms similar to motion sickness. As a result, simulator sickness is another important factor to consider when designing a VR user experience:

- *Simulator sickness* "is an uncomfortable side effect experienced by users of immersive interfaces commonly used for VR. It is associated with symptoms such as nausea, postural instability, disorientation, headaches, eyestrain, and tiredness [154]."

In principle, simulator sickness is induced by a sensory mismatch between visual stimuli and appropriate vestibular or proprioceptive signals. However, there are other factors that can cause motion sickness, particularly those caused by the VR application's hardware, such as latency, flickering, and frame rate. Additionally, those factors can be related to the application, such as how accelerations work in scenes, visual flow, direction or angular presentations, and how flickering is displayed [155–157]. Lately, significant advancements in both the hardware and software of HMDs have been made, which has resulted in a reduction in the occurrence of simulator sickness but has not eliminated it entirely. Moreover, in addition to hardware and software, network parameters have an important role if a VR application is configured to communicate through a network.

Understanding the role of network latency in simulator sickness is important since VR experiences can be severely degraded or made really obsolete due to network delays. As a whole, virtual reality (VR) systems take advantage of reducing latency, which leads to increased usability and safety.

However, there are no hard and fast rules on how long a delay may be in VR multiplayer games. Values observed in research range from 50 to 210 milliseconds . An example is a VR learning application that investigated the impact of delay on motor skills, and interaction [158]. Delays of more than 75 milliseconds are disruptive to motor performance but have little influence on a person's feeling of control or ownership, which are damaged by delays of 125–210 milliseconds. However, this study is limited to how a single player's body moves and how the user interprets that player's body. On the other hand, the research proposed that characters in computer games should be moved in fewer than 150 milliseconds [159], noting that a delay of 50–90 milliseconds lowers the impression of being there in a virtual environment [160].

2.2.5 Readability

Visual information must be created and shown clearly and pleasantly to provide a positive user experience while reading text on a screen [161]. The capacity to read sentences from a given source independent of their meaning is critical for readability. *Readability* as a term is used within different domains so that it can be referred to as the accuracy of reading [162]; as the ability to understand the text with taking into account the speed of reading [163]; or to the visual representation of the characters themselves [164]. In any one of these scenarios, it is necessary for the reader to comprehend the content. Because there are no screen size constraints in the UI for VR, this provides an advantage in terms of the ability to present information in all directions [5]. However, it is important to display information where items are in relation to the user so that information is easily comprehended. When it comes to showing text in VR, not only the font size but also the distance between the user and the text are important considerations. As a result, Google developed a new font size unit called *dmm*, which stands for *distance independent millimeter*, defined as the height of the character of 1 millimeter on the distance of 1 meter [165]. That was suggested since there was no standard measure for angular size in VR at the time, providing the same view from any distance on every screen. As a result, the unit soon received attention and is currently applied in both industry and research. One study on readability in VR looked at doing everyday tasks in VR. As with a web browser or emails, they produced static graphics and requested users to view and engage with them. Reading speed and accuracy were also tested [166], and the results demonstrate that continuous head movement improves reading in VR. Another research reported in dmm looked at fundamental characteristics such as text box size, font size, and preferences [167]. The findings show that the text size should be 41 dmm $+/-$ 14 dmm. Participants also preferred white text on the black background and sans-serif typeface. However, in previous studies, participants were not allowed to change font properties to determine their preferred settings.

Contrast is another important factor to consider when determining whether or not the text is easy to read [168]. If there is a good contrast between the background and text colors, readability will improve for all users, not just those with low or no vision. Text with effective contrasts is easier on the eyes, scans faster, and is more readable overall. Poor contrast, on the other hand, will make reading more unpleasant and time-consuming. Even though it is not made in particular only for VR content, *The Web Content Accessibility Guidelines 2.1 (WCAG2.1)* is one of the most widely used standards for ensuring that the content of websites is accessible [169]. The underlying concept behind this standardization is to ensure that website visitors can easily understand, navigate, and interact with them. An indication for the contrast ratio between the color of the text and the color of the background varies depending on the size of the text. Large text is defined as text that is at least 18 points in size or 14 points bold. Here are suggestions that are proposed by WCAG2.1 [170]:

- Level A standards ensure that all users can access the information on the website and that an alternate version is also provided.
- Level AA requires a contrast ratio of at least 4.5:1 for standard text and a contrast ratio of at least 3:1 for large text.
- Level AAA requires a contrast ratio of at least 7:1 for standard text, while a contrast ratio of 4.5:1 is needed for large text.

However, standardization of accessibility guidelines, particularly for VR applications and solutions, is not yet available.

2.2.6 Social Acceptability

Even though VR technology has gained in popularity in recent years, it is still rather rare in everyday situations other than home usage. As VR devices become all-in-one solutions (requiring no additional gear), this technology has the potential to be utilized everywhere. Nonetheless, the presence of unnoticed spectators may have an influence on UX and raise concerns about the social acceptability of this technology:

- *Social acceptability* refers to a prospective judgment toward a technology or measures to be introduced in the future [171].

Investigating acceptability in general and social acceptability, in particular, is critical for predicting future success and adoption rates for any new technology. Another difficulty in dealing with social acceptability in VR is the lack of accessible, standardized measures. The Social Acceptability Questionnaire (SAQ) is a nonstandard questionnaire designed to measure social acceptability from the performer's perspective [172]. It is made up of 11 questions evaluated on a 7-point Likert scale, containing statements about Public VR, Users, Public Communication, Interaction, Isolation, Privacy, and Safety.

As of yet, little study has been conducted to determine how successfully VR technology will integrate into society. Nonetheless, as the following research from recent years demonstrate, interest in the area is growing. According to research, [173], what is socially acceptable from the perspective of a spectator is highly dependent on the context and setting. VR seems to be appropriate for usage in areas such as a bedroom, the metro, or a train, but not in settings where people are expected to converse, such as living rooms or public cafés. This study provided the framework for future research, but it still faces the limits of an online survey in which participants do not get to use the technology they are rating in person. Another research compared how users felt in three situations: with and without physical boundaries and in an empty room [174]. This study concentrated on the physical environment's configuration, but it also did not consider the emotional and interactive influences on the experience.

2.3 Influencing Factors

Many different forms of human-computer interfaces have been produced for a wide range of general and specialized scenarios, including those in which users have gained competence with the innovations through use, and these interfaces have been designed for a number of various technologies [175]. UX research is primarily concerned with enhancing system design and evaluation, as well as supporting the emotional needs and objectives of users [176]. For the purpose of gaining a deeper understanding of the concept of UX, a variety of distinct frameworks and models have been established [176–178] incorporating a range of different components and measures for UX [177, 179]. These frameworks are provided in different ways, with the goal of comprehending and documenting as many types of experiences as possible created by interacting with the system [175]. Studies of UX do not only concentrate on task-related characteristics, but they also investigate emotive qualities, feelings, meaning, and the value of interactive products and services [178]. The definition of an influencing factor is given as:

- *Influencing Factor (IF)* is any characteristic of a user, system, service, application, or context whose actual state or setting may have influence on the Quality of Experience for the user [180].

From the definition, it is stated that many different factors may have an effect on the experience of a user. Both UX and QoE report on three main categories that have an influence: user, system, and context. When evaluating the experience of a game, it is necessary to know about these factors in the research. They are either being evaluated or need to be controlled in order to get reliable and valid results from experimental user testing. In addition to this, these factors are necessary for the construction of models for UX or QoE. However, the main categories should be further divided into specific aspects, which will be discussed in the next section. Furthermore, each factor is connected to a unique set of characteristics and features that have an effect on UX as a whole. Meaning that they can expand the capabilities of UX, but at the same time, they make it diverse as a wide variety of concepts for different disciplines and technologies have to be incorporated. The research literature specifies just a few UX models for immersive virtual environments [181]. Those UX models are rather focused on some components, such as presence and immersion, rather than the full overview. However, the User eXperience in Immersive Virtual Environment Model (UXIVE) was built based on four distinct UX models [107] and gives as a result the overview of key components of UX for Immersive Virtual Environment (IVE), as it has been presented in Sect. 2.2. Despite the relevance for UX, overall less focus is given to factors influencing UX, regardless of the fact that they can improve measuring and modeling of UX [175, 182–184].

2.3.1 *Influencing Factors for Gaming*

In recent years, there has been an attempt to explore UX in terms of the basic concepts and action plans for creating design changes in order to standardize playtesting across video game creation in general [185, 186]. However, the video game and VR industries, as well as researchers, do not yet have guidelines on how to arrange the different methods to quantify the gaming experience in VR into a framework. Each player's experience is especially important in serious gaming, where the objective is to help individuals learn and modify their behavior. The sort of evaluative UX measure used varies depending on the type of serious game being evaluated, but effective measures are the most often employed to determine how effectively game-based learning applications or games for sports and health make people want to play them [187]. Despite the fact that these tests look at how individuals feel, it is believed that positive emotions have something to do with how effectively the brain can store and recall information [187].

All games are defined by what the participants do and how they make decisions since game players are not passive, and it has been shown that the game is entertaining because players take on roles [188]. Further on, players define their own opinions based on their choices, and their actions and situations allow them to believe in their own ideas and gain experience and knowledge. Serious games allow individuals and groups to collaborate to create a wide range of situations and strategies that can be tried and assessed. This is accomplished by immersing users in a virtual environment that mimics a real-life scenario. The findings of a study on serious gaming and UX of learning demonstrate the benefits of using serious game elements in a context of user-centered design and co-creation [188]. Benefits include, but are not limited to, aspects such as sociability (the ability to interact with other people), decision-making, exploration, immersion, and overcoming distance. However, the relationship between those factors and how they influence the users' overall experience is not reported. When it comes to the standardization of the influencing factors for gaming, research from QoE provides a framework and taxonomy for QoE aspects while still maintaining and building on top of the hedonic and pragmatic UX aspects.

Quality Aspects are defined as individual categories of the quality of the service under investigation [180]. They each include one or more quality features. Those features and factors that are important for gaming are put in a logical order to show how they depend on each other. At the top of the taxonomy, quality factors are divided into user, system, and context. Overall, the taxonomy does not show all significant links between factors or aspects, and interactions across layers are not shown; however, certain taxonomy users may be especially interested in such cross-layer linkages.

Building on top of it, the Recommendation ITU-T Rec. G.1032 [189] was written with the idea to tell gaming service providers and evaluation experts about things that could affect gaming quality of experience. In general, the recommendation gives a helpful overview based on [180], with the expansion to include cloud and online gaming services. The recommendation also presents how IFs work in a passive watching-and-listening scenario, an interactive online gaming scenario, or an interactive cloud gaming scenario. The criteria discussed in the recommendation mostly target typical video game experiences, which may or may not be complemented by audio. However, other interaction modalities have yet to be addressed. Furthermore, the recommendation only gives a basic description of devices used for interaction in AR or VR technologies.

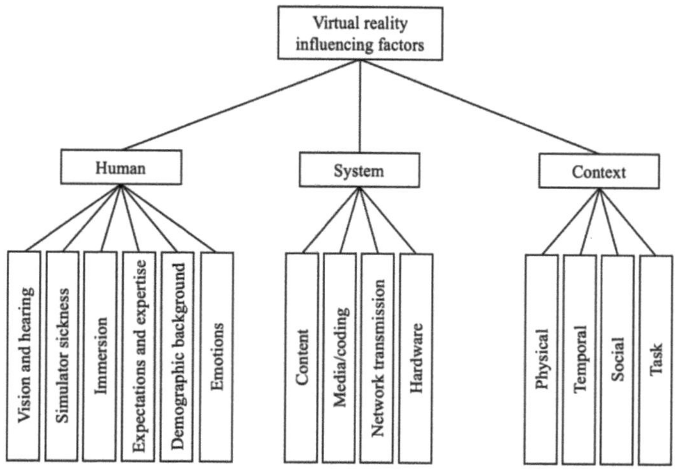

Fig. 2.4 Virtual reality QoE influencing factor categories [15]

2.3.2 Influencing Factors for VR Services

The Recommendation ITU-T Rec. G.1035 [15] as shown in Fig. 2.4 categorizes and analyzes the characteristics that impact the experience of a VR service in order to provide a tool for the identification of methodologies that may be used to assess VR experiences. When compared to traditional video and audio, the multimodal experience of VR imposes a new set of criteria for the evaluation of experience, with the challenge to assess VR as having real-life immersive imagery and sounds, in addition to interactivity [15]. Because VR technologies are still in the stages of development, this Recommendation focuses primarily on omnidirectional video services and provides an overview of influencing factors over three main categories.

2.3.2.1 System

System influencing factors "are related to media capture, coding, transmission, storage, rendering, and reproduction/display, as well as to the communication of information itself from content production to user" [180]. They take into account the way hardware and software are built and designed, as well as the game that is supposed to be played:

- *Hardware-related factors*: Physical and technical characteristics of VR devices, such as portability, screen type and size, headphones, input controllers and tracking systems, resolution, viewing angle, and distance, are all considered hardware factors. Furthermore, HMD factors like weight, size, and device heat impact the wearing comfort and overall experience. Regarding the field of view, a user who has a wider view is more likely to feel as if they are immersed in the experience. Movement monitoring is also important, and it can be based on two technologies: outside-inside and inside-outside. The outside-inside technology is the one where additional motion-sensing sensors are used, and they track the position of HMD from the outside, while the inside-outside is using only the HMD as the sensor to detect the position of the head or body part relative to the surroundings with cameras on the HMD itself.

- *Network/transmission-related factors*: Only online VR services have to consider network and transmission-related factors. This section addresses factors like delay, bandwidth, and packet loss. Remote cloud servers for VR services are often used to offload work from HMD and outsource more demanding computational activities to the cloud, although such solutions may cause extra communication delays. All delays up to 50 milliseconds are indicated to have a minimum influence on the experience, whereas delays higher than 100 milliseconds considerably degrade it.
- *Media/codec related factors*: Media compression and the characteristics of network transmissions (such as jitter, bandwidth, packet loss, and rejection) are the main points. However, some special points are pointed out in comparison to 2D services, such as that for video, it is recommended to use different resolutions depending on whether the image is sent and displayed within the user's field of view. This technique is known as viewport-adaptive or tile-based omnidirectional video streaming. Furthermore, because the pixels are in a 360-degree sphere surrounding the viewer, only around 25% of the total pixels of the full resolution can be observed by the user, depending on the device utilized and the field of view. Also, the important difference between system characteristics is the speed with which images are presented when compared to the recommendations of 2D video services because simulator sickness can be caused by slowness in showing images in relation to user motions.
- *Content-related factors*: VR content is integral to the experience, and it has distinct requirements from other forms of multimedia content. Aside from having great video and audio, VR content needs to be stitched, have special effects, be in stereoscopic 3D, and be put together well. Despite the fact that there is no universal way to classify games, the recommendation highlights important factors such as the game's genre, mechanics, and rules; the temporal and spatial accuracy required to play well; the temporal and spatial complexity of video content; game speed; visual perspective; aesthetics and design features; and learning difficulties. Most of these factors are still not categorized because of a lack of specific definitions and standards.

2.3.2.2 Context

Context influence factors "are factors that embrace any situational property to describe the user's environment in terms of physical, temporal, social, economic, task, and technical characteristics" [180]. Overall, the recommendation is based on a number of different factors, such as the physical environment, the temporal context, the established social context, and other service factors, such as the task, the availability, and the cost or goals connected to the usage of a system:

- *Physical context*: Physical context aspects are those relating to the user's surroundings when using VR services. Depending on how the experience is, factors such as lighting and temperature might have a bigger or smaller impact. The unique aspect of VR that uses HMD is that proper security measures such as virtual barriers or transparent mode are required to make the user aware of their physical surroundings if necessary. Otherwise, another individual who is not participating in the VR experience can observe the user's movements.
- *Temporal context*: The length of usage and the frequency of usage are both examples of temporal context factors. It is advised that users who use VR equipment do not do so for lengthy periods of time.
- *Social context*: Factors of the social context include the different ways in which VR services are used, such as individually or in multiplayer mode. Sharing and utilizing VR services together might result in varied experiences depending on whether the individuals around the user are related or unknown to the user.
- *Task context*: The goals for which a user utilizes VR services are examples of task context influencing factors. Users could concentrate on various aspects of the experience depending on whether the purpose of the game is for entertainment or the VR service is used in a serious context.

2.3.2.3 Human

Human influence factor is "any variant or invariant property or characteristic of a human user. The characteristic can describe the demographic and socioeconomic background, the physical and mental constitution, or the user's emotional state" [180]. The user's past experience with VR, particular game or genre, intrinsic or extrinsic motivation, demographics, and dynamic aspects like emotions, as well as the user's ability to see and hear, are all included as factors in the recommendation:

- *Vision and hearing*: A user's ability to enjoy VR might be impacted by vision or hearing issues. Although different types of impairments are possible and occur across the population, loss of eyesight or sensitivity at high frequencies are frequent impairments connected with user age. The most frequent visual impairments may be corrected using lenses or glasses. However, it should be noted that not all HMDs can be used with glasses beneath.
- *Simulator sickness*: Simulator illness may be caused by a variety of variables, not all of which are technical. Simulator sickness is most often caused by numerous covariates, including the user's age and gender, and depending on them, greater or lesser symptoms may arise.
- *Immersion*: Regardless of VR technology, some people are more likely to feel immersed than others. The amount of immersion that a person may feel is determined not just by the VR system but also by the individual's competence

and the feasibility of the experience itself. It is unclear how individuals develop a greater or lesser tendency for immersion, and it is up to future research.
- *Expectations and expertise*: Depending on the user's experience and whether or not he has used a similar system before, his expectations may change. Those who use a system for the first time may be surprised, while those who have used it before may focus on tasks and details.
- *Demographic background*: The term demographics refers to statistical representations of variables such as age and gender, in addition to socioeconomic information such as employment, education, income, race, marital status, and other similar factors.
- *Emotions*: Emotions have the ability to affect the experience in general, including VR services, in the same way that the stimulus activates the receiver's emotional reaction, or the meaning of a stimulus might lead the recipient to have an emotional response to it. The increasing accessibility of VR equipment and technological breakthroughs have inspired new uses, one of which is the study of human emotional states.

2.3.3 Methods Considered for Research Objectives

Following on the work presented with ITU recommendations for influencing factors on experience for gaming and VR services separately, this research aims to explore what factors are influencing the experiences of users for VR serious gaming and their connections to the UXIVE model, as shown in Fig. 2.5. Although there are studies that show the impact of individual factors on user experience, a systematic review of all such papers would be too complicated due to differences in experimental and method designs, and the results would possibly be unreliable due to incompatibility. It should be noted that the focus of this research is not on detailed research of new methods of evaluating VR or serious game experiences, nor on the research of new technical possibilities of VR technology and network data transmission. Therefore, this research uses VR devices available on the market, focusing on HMD and existing research methods. While some interesting findings and methods for behavioral and psycho-physiological assessments are currently available, such research is still in its early stages. It is not yet reliable enough to measure all of the specific characteristics of VR experiences. Traditional methods such as questionnaires, although not originally developed for VR technology, are focused on only one aspect and can be combined regardless of technology.

As this research focuses on VR technology that uses HMD, the user cannot see the real world while playing. As a result, according to the recommendation on physical context, it is necessary to ensure that the moderator observes the participant during the experiment in order to avoid injuring the participants. As a result, all experiments are conducted in the laboratory, allowing for safe operation while the moderator monitors movement and warns participants. Given the choice of laboratories as a space with the focus of research on VR technologies available on

Fig. 2.5 Overview of IF for VR services with focus on investigated subfactors

the market, certain influencing factors could not be part of this work and research. Specifically, concerning the system, this research will not use different media and codec factors but will use those that a particular HMD originally allows. Also, given the specificity of the human factors, tendency-based parameters defined with ITU-T Rec. G.1035 [15] such as emotions, immersion, and simulator sickness were not additionally pre-monitored before the study to know the tenancy of each user, but only parameters were just monitored as consequences. However, participants in the experiment could report discomfort at any time, and in such cases, the experiment would be terminated immediately without any consequences for the participant. Participants were informed about this, but there were no such situations during this research. Furthermore, given the recommendations that VR should not be used for an extended period of time, the temporal context was also skipped, and all experiments were set within the recommended time limits for a particular VR device. Since users were aware that they were coming to participate in the experiment, the task context was as well always the same. However, to the greatest extent possible, all of these variables were used as control variables in the study.

2.3.3.1 Research Questions

When it comes to VR technology and Hardware IF, several studies have been conducted to investigate the difference between HMDs and CAVEs, with varying findings. While the CAVE environment was compared to interaction in the real world [190], it was observed that users took longer to perform tasks, and they made more errors when participating in the CAVE environment. In another study [191], researchers evaluated anxiety in both the HMD and the CAVE settings. According to the findings, the level of anxiety brought on by a phobia of heights was significantly increased, while participants were in a CAVE. Other studies investigated HMD solutions as opposed to a tablet. One tested how seeing various emotional video content on an HMD or a tablet altered the viewer's environmental attitude and behavior [192]. The study's findings demonstrated a significant positive association between using an HMD and self-reported measures of greater immersion and higher emotional impact.

The concept of gamification emerged in the gaming industry as a response to the opportunity to make use of the intrinsic motivation that games provide. Gamification was developed with the intention of engaging people in non-entertainment areas such as education, simulations, health care, and sports. Peng et al. [193] investigated an exergame by modifying game features, such as difficulty adjustments or dialogues, with the intention of influencing the basic needs, autonomy, and competency individually in order to predict enjoyment and motivation. This research was conducted in accordance with the SDT methodology. They found that a rise in both players' requirements for autonomy and competence improved their level of involvement in the game.

Still, when it comes to VR, it is still mostly unclear how it is at the moment used on the market to influence experiences of VR serious games. Therefore, research focusing on an overview of motivation in VR exergaming is pointing out that it is necessary to thoughtfully incorporate exercise and motivation-related theories into VR research as well as to give careful consideration to VR definitions, together with software possibilities [194]. Therefore, in order to explore influence of Hardware IF, an RQ is stated as:

RQ1.1 Can the use of a virtual reality set-up increase the users' motivation and engagement for VR serious games?

In addition, when it comes to System IFs, VR serious games, just like any other classic video game, may be designed as either single-player or multiplayer experiences, with synchronization of the gameplay happening via the Internet. In order to investigate the influence of the delay on motor performance and interaction in VR learning, a study was conducted [158]. According to the findings of that research, delays longer than 75 milliseconds have a negative impact on motor performance, but they have no impact on perceived agency and ownership, which are influenced by delays lasting between 125 and 210 milliseconds.

On the other hand, that research concentrates only on the motion lag experienced by a single player and the impression of the player's body alone. However, there are no set guidelines for the maximum amount of delay that may be used in VR multiplayer exercise games. When it comes to computer games, there has been more research done, and according to the study, the recommended value for managing characters in computer games is less than 150 milliseconds [159]. Nevertheless, it was found that the sensation of presence is diminished in virtual situations already for delay levels ranging from 50 to 90 milliseconds [160]. Therefore, in order to investigate influence of network transmission IF for VR serious games, an RQ has been made as:

RQ1.2 Does the network delay during VR serious games have an influence on the user experience, in particular regarding the perception of own delay?

Concerning Content IFs, the feeling of presence and immersion are critical when it comes to simulations in VR, and the importance of it has been shown in research [37, 195]. This sensation may be disrupted by any touch with the actual environment, such as using controls or sensors [72]. Because of this, new HMDs are enabling features that allow users to engage through hand tracking rather than using controllers. This approach has the potential to lead to the development of highly captivating use cases in VR gaming, as well as in other domains where VR is used for serious games. Learning already makes extensive use of VR [73], particularly in the fields of medicine [74] and manufacturing [75], as these are the kinds of disciplines that have been linked with accuracy and hand motions play a significant role. In a VR simulation, there is research that reports on hand tracking and visualization. In this simulation, a design that offers to monitor the fingers and palm has been built [79]. The emphasis of the research was on studies in which participants were required to duplicate certain hand postures, and comments on performance were based on how quickly the tasks were completed. However, in the field of gaming, research on virtual environments [77] has focused on particular algorithms for how to achieve precise hand tracking via various sensors [78, 196]. However, besides integrating sensors, it is also important to ensure a good user experience for VR serious games to achieve good simulations and results.

Based on the recent release of the build-in hand tracking technology and the fact that there are still little to no scientific studies performed to test for influences on user experience, the RQ about Supported Interactions IF is formed as:

RQ2.1 Does interaction ability and generated content in a virtual environment have an impact on the user experience in VR serious games, in particular on the perception of flow and presence of players?

Moreover, when considering content as IF, it is interesting to observe research done for UI of VR and Environmental Design IF. Already mentioned recommendations are made in general for VR interfaces (Sect. 2.1.2.4), but when it comes to research of UI in VR serious games, there is a study that is focused on designing for VR exergames [68]. According to the findings of this research, the player viewpoint in exergames may typically be set to either the third-person or the first-person perspective. If a game is played from a third-person viewpoint or uses a virtual body user interface, the player will see a digital version of themselves, an avatar [69]. Because the user sees the game from a first-person viewpoint, they do not see their bodies but rather just the view. Because the vision in VR matches the location of the player's head, the first-person viewpoint is a natural fit for VR and should be used wherever possible. In addition, it has been demonstrated that people had no trouble comprehending the first-person viewpoint while using the HMD. When it comes to the research of video games and VR they are comparable [197]; as a result, user interface research concerning video games may be applicable to VR. On the other hand, when compared to the implementation and the way information is represented in the non-gaming environment, there are some significant changes in games. However, rather than focusing on the content of the UI, other VR studies are more concerned with the difficulties associated with using immersive technologies, like parameters of motion sickness [156] and physical feedback [158]. Therefore, with the idea of exploring difference for the content design of UI for VR serious games, the RQ was formed as follows:

RQ2.2 How do the differences in complexity and position of user interface elements influence user experience in VR serious games?

As per Context IFs, as part of this work, only Social IF will be considered; when it comes to VR in general, there is still little research on it, particularly on the social acceptability of VR technologies. However, very recently, there has been an increasing interest in the topic, as some studies are showing. According to the findings of a study [173], social acceptability, as seen from the observer's viewpoint, is strongly dependent on the scenario and the environment. Building on that, in a different study on the social acceptability of input modalities for HMD display [198], it is argued that from the users' perspective, considering a social context that is appropriate to the use of the technology is mostly related to their mental construct about accepting or rejecting a specific gesture in that setting. Despite this, the scope of this research was limited to AR technology, which does not have major challenges such as those connected to the conventional VR characteristic of reduced vision of the real world. Further on, research has made the goal to understand and identify the aspects that influence the experience of the performance as well as the experience of the spectators [199].

According to the results, using a VR headset in a public setting such as a university does not disturb or make the viewers uncomfortable. Still, because of the complete overlay of vision provided by VR headsets, one of the most significant worries expressed by users was connected to potential threats to their safety, such as the fear of accidentally colliding with real-world items or other people. Another study [200] is an example of an effort to find a solution to this problem that was focused on tools for 3D scene reconstruction in VR environments to increase obstacle awareness and enhance interactivity. In addition, similar to the findings of previous investigations, neither the spatial position nor the environment is subject to any control or modification. As a consequence, it is necessary to determine whether or not the resultant criteria of UX still apply to various situations. So, the RQ was formed in accordance as:

RQ3.1 Does the social environment influence user experience, resulting in the social acceptability of VR serious games?

When considering the social components of games and motivation, it is necessary to highlight the importance of studying multiplayer games. Users have different experiences when they compete against other people as opposed to when they compete against themselves in a game. According to research [201], participants report experiencing more enjoyment when participating in multiplayer gaming scenarios. Additionally, via communication between the participants, more information may be exchanged, which may contribute to building a stronger level of trust between the players [202]. When it comes to VR, research on a conversation is mostly connected to VR teleconferences [203], but not on having it as an additional feature to the main game play.

It has not been well examined how VR affects players' motivation and participation in multiplayer VR serious games, particularly when there are ways for participants to communicate with one another. As a result, the RQ for this work has been formulated as follows:

RQ3.2 How does the opportunity to have conversations in multiplayer VR serious games influence social presence?

When it comes to the Human IFs, user characteristics like as age, gender, and previous experience have been factors that have been taken into account already from computer applications [204]. Incorporating such factors to understand the behaviors of certain users with certain technology is not uncommon in research. Outside of VR research, it has been observed that gender and age impact the perception of multimodal human-machine interactions [205]. When it came to judgments of how simple it was to get the information, factors such as age and gender had a major role. Tasks were perceived as difficult by a group of participants of an older age, which suggests that they may have to put in more effort than participants of a younger age group to solve those tasks. In addition, users of various ages have varying preferences regarding online communication [206].

More recently, some research with a similar objective has also been carried out in VR with an emphasis on cybersickness and presence. Results have demonstrated that there is a strong influence on 3D video material. Female participants have reported feeling more present and less nauseous in the virtual world [207]. In addition, researchers are recommending varied durations of VR exposure according to age and gender when it comes to different applications in VR [208]. Accordingly, for VR serious games, the same trend could be assumed, so the RQ about it has been created followingly:

RQ4.1 How do age and gender affect the perceived quality and user experience for VR serious games?

Similarly, it has been shown in other fields that prior experience working with technology is an important factor. When developing an application, it is important to make an effort to deliver a solution that is highly responsive to the requirements of the user. This is particularly important when adopting the principles of user-centered design [13]. The experience will increase according to the degree to which this approach can be effectively integrated into the program [14].

There are theories that users like games in which they can play successfully but have not yet mastered [112, 138]. The difference that exists between gamers, casual gamers, and non-gamers makes it more difficult to maintain the appropriate level of equilibrium inside a game. Therefore, application designers and developers are suggested to understand their users' behavioral capabilities.

However, VR, in particular, has for most users still the so-called wow effect [209], where users who are faced with VR for the first time will report having more positive emotions compared to those who have tried the VR technology before. So, building on top of that effect, the RQ regarding the influence of previous experience with VR for serious games is stated as follows:

RQ4.2 Does the previous experience with VR have an influence on the current user experience within VR serious games?

2.4 Summary

This chapter provided an introduction to the theoretical overview of the main concepts and related work regarding UX, serious games, and VR. In order to cover all three aspects, they were discussed separately and in relationship with each other. VR technology has been introduced, including theoretical points and a summary of recent on-market hardware solutions. Further on, several important aspects related to VR have been defined and discussed, such as simulator sickness, 360-degree UI, and interactivity. Later, those are addressed as important influencing factors for VR services.

Even though, in recent years, VR has entered the so-called second wave of its popularity, some definitions and theoretical backgrounds are not new and have been introduced with the first devices. However, with the development in hardware, the literature was revising and developing new definitions, which led to the fact that the term VR can be used with different meanings for different forms. Therefore, this work was built on some common theoretical basis, starting from the Reality-Virtuality continuum to distinguish real and virtual environments over AR, AV, MR, and VR technology. Some of the given definitions are more or less specific to the other, and some include the type of hardware, while the others are more focused on visual medium or on the provided level of immersion. Still, as there was no standardized definition yet, this work will be based on the recommended definition given by ITU for VR. In particular, this definition is used as the base for determining IFs for VR services in ITU-T Rec. G. 1035.

Similar challenges are presented with UX definitions, as the term has been developing together with HCI research. Still, as there is a standardized definition by ISO 9231-210 standard, this work is building on top of it with the understanding that the definition still leaves some space for interpretation. When it comes to the UX in VR and overall models, research for it has been limited and not specific but rather incorporating VR into the Immersive Virtual Environment Model. However, that model provides ten main components and relationships to each other. Similarly, assessment methods for each of the main components of UX are discussed, focusing on subjective quantitative methods, as those will make it possible to study each factor in the set-up of user studies.

An overview of the state of the art for influencing factors on experience for gaming and VR services is presented, and based on work from ITU and combining the ITU-T Rec. G.1032 for gaming and ITU-T Rec. G.1035 for VR services influencing factors for the system, context, and human are named. However, those recommendations are not focused on VR serious gaming and are not reporting on relationships to UXIVE model. Therefore the work of this thesis is focused on the aim of exploring connections based on in-lab user studies between influencing factors and UX for VR serious gaming. Finally, the RQs for this work are formed and provided based on the existing research gap of design principles and recommendations for UX of VR serious gaming.

Chapter 3
Methods

3.1 Overview

The methodology that was used for this research is presented in the overview based on the selected influencing subfactors and related research questions that have been the main focus of this work. As part of it, several different VR applications, each centered on a different set of features, were developed with the goal of investigating UX in VR serious gaming through empirical investigations done in a laboratory.

The main difference between the applications is in the field of usage together with the elements that were in focus for the particular study. As applications are made to represent possible VR serious games, but with additional features and possibilities for adjustments that on-market solutions do not have, each application was tailored for a particular user study. Accordingly, this chapter will first present each of the five developed applications with the set-up, ranging across different system and content features. Still, every application included at least one interactive scene where the user could interact with a virtual world using a controller, leaving 360-degree videos used only in scenes to enhance virtual world features.

Regarding methodology, after presenting the developed VR applications, the motivation based on research questions and a connection between the study and the application itself will be provided for each user study, including an explanation of the procedure, used conditions, and changed parameters between the conditions and participants that took part in the study.

An illustration of an overview and connections between the influencing factors and subfactors to corresponding user studies and used VR serious game applications is given in Fig. 3.1.

Serious games are often used in activities that can be extremely monotonous or tedious, for example, physical activity, cognitive training, and rehabilitation exercises, as a complementary tool to increase user engagement [210] (as explained in Sect. 2.1.3.1). Interactive and digitally connected technologies have great potential to support recreation and social connections, both of which are important aspects

T. Kojić, *User Experience for Serious Games in Virtual Reality*, T-Labs Series in Telecommunication Services, https://doi.org/10.1007/978-3-031-75530-9_3

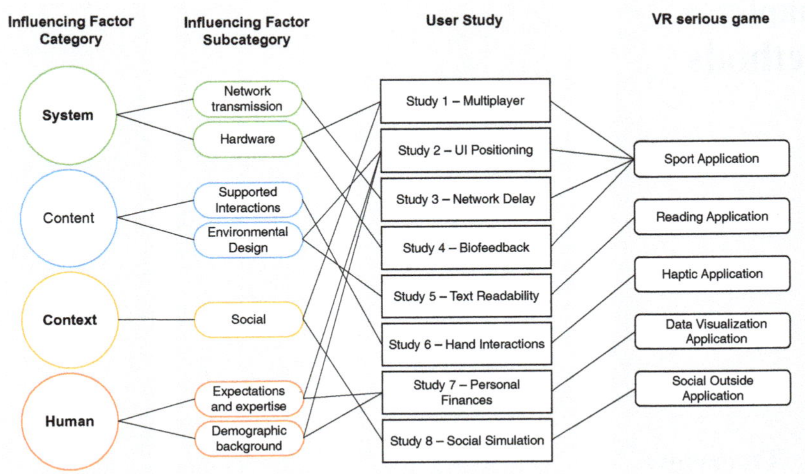

Fig. 3.1 Overview of the researched IF in connection to the conducted User Study using a VR particular serious game for it

of promoting a sense of well-being and improving quality of life [68, 211, 212]. Employing the multimodal set-up of the system, sensors, trackers, and cameras, it is possible to use parts of the body other than the hands and fingers while playing serious games, thus encouraging full-body exercise while allowing the player to have fun. As already mentioned, this genre of games is more often called exergames. In addition to supporting physical health, these technologies also have enormous potential in the process of creating a general sense of well-being. As a result, the concept of exergaming was chosen as the basis for the Sports application.

Serious games, on the other hand, do not have to be exercise-based; thus, the other applications were designed with more static playtime in mind. As the way people work in the office changes, solutions are focusing on the potential of bringing serious games to the workplace in a variety of fields. More people are working from home, more people desire more flexible work options, and more people are reconsidering what it means to work in an office. The ability to work inside and outside of traditional business environments and during normal business hours is referred to as a "virtual office" [213]. Although there are no physical boundaries, the virtual office is still connected to real places. Such offices can be located in various private or public places [214] and enable individuals to perform work at any time, including when the traditional working day has ended. However, in order to facilitate focusing on specific aspects, in this chapter, the applications are further divided and are given appropriate descriptive names while adapting to specific RQs within specific studies. Also, unlike the exergame, which placed primary emphasis on the overall game, the following applications used game elements solely for the purpose of simulating activities typical of an office environment, such as reading, understanding, and handling.

With regard to the social setting, the purpose of this work is to make a contribution to the improvement of knowledge and comprehension of the elements that influence the adoption of VR technology in social settings. The goal is to overcome restrictions and address some gaps discovered in past studies such as [173, 174] (from Sect. 2.2.6). Therefore, the additional application having both dynamic and static interactivity of gameplay has been created with the scale of social environments depending on the number of people present and their proximity.

3.2 Set-Ups

The following subsection provides an overview of developed VR applications with the intention of explaining in detail each set-up solution, including examples of UI visualizations in VR and the hardware used for it.

3.2.1 Sports Application

With the aim to investigate how to build and develop VR exergames, the Sport Application was created specifically to be able to adjust the settings required to answer research questions. Rowing was chosen as the sport for the VR exergame, and in this set-up, a rowing ergometer was connected to the Sport Application, which contained several scenarios. The main set-up of the system for the Sport Application is the rowing ergometer, personal computer (PC), and HMD. In particular, the application was built to work with the Augletics Eight ergometer, as well as with the Concept2 Model D rowing with a smartphone connection for the ergometer. Those ergometers were chosen as they have the possibility to be connected to the PC, and real-time data from the ergometer, such as distance and speed, could be transferred wireless. On the PC, there has been installed a Unity application able to get transferred data and display them within the Sport Application. For displaying the data, an HTC Vive set was used as HMD that is connected to the PC. Additionally, the BreathZpot sensor could be added to the set-up to measure the participant's breathing. Also, in order to enable communication between players, a wireless in-ear headset can be added to this set-up. An illustration of the basic set-up for the Sport Application is given in Fig. 3.2.

Overall, for the needed solution, there have been created in four different scenes.

The main virtual scene was created using a Unity project that had a simulation of a lake that was surrounded by mountains on a sunny day. From inside the rowing scull, the player had a clear view of the surroundings. The first scene of this application aimed to represent different statistical data during gameplay. The information that is presented includes the boat's speed, the distance it has gone (from its starting place to its current location), and the amount of time that has passed (from the beginning of the race). Depending on the setting, there were

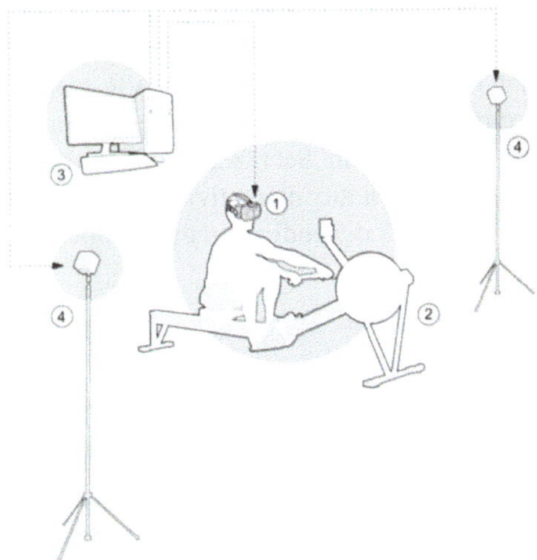

Fig. 3.2 Set-up of the Sports Application where numbers on the illustration represent: 1. HTC Vive as VR HMD set, 2. Rowing ergometer, 3. PC with Unity application, 4. HTC Vive base stations

Fig. 3.3 Players view on cockpit at the front of the rowing scull with gamified metrics visualization

different visualizations of data possible. The difference was in the amount of information complexity (how simple or complex the data are represented) and in the placement of the information (displaying statistics closer or further away from the player). As can be seen in Fig. 3.3, these statistics were given in a gamified format in order to simplify their presentation and make them simpler and less difficult to comprehend. In order to make the problem more complex, certain extra data were included, including a continuous power reflection (power curve) that was represented as strokes per minute and an estimated energy consumption measured in watts. The complex representation is shown in Fig. 3.4, which solely uses numerical representations.

In order to define the position where to display statistical data (simple or complex) in the exergame, Bowman's classifications [215] were used. Heads-up display (HUD) are often seen in the four corners of video games, but for VR scenes, these positions cannot be used for the display of information since they are outside

Fig. 3.4 Players view on a coach boat that follows the player with digital visualization of metrics with a countdown screen for the beginning of the workout

of the recommended comfortable zones for putting objects in a VR [61]. However, another representation for HUD is to take the shape of a cockpit. A user interface formed as a cockpit, according to Bowman, may be added to a simulation game to show information such as speed. Therefore, a cockpit visualization was chosen to serve as one of the two locations for statistical data in this application, as illustrated in Fig. 3.3. Additionally, another location for gathering statistical information was designed to be located somewhat further away from a player than a cockpit in order to provide integration. The approach that Llanos and Jørgensen took to the UI problem as an integration with the surrounding environment was followed by the design [216]. The objective was to create a scenario in which statistical data could be presented in a manner that appeared natural without deviating from the conditions that had previously been established. Because of this, a coaching boat was developed, including an option to display all of the statistical data that had previously been arranged for the metrics. The rowing scull that the player was in was being followed by a coach boat, just as it would be in a real training session, as it is shown in Fig. 3.4. It kept pace with the player and maintained a constant distance from them in order to ensure that the data could be read and observed without difficulty at any point during the race.

Secondly, the infrastructure required to support multiplayer modes has been incorporated into this application. The virtual environment was created as a Unity app as part of the multiplayer settings, and it showed two red sculls sailing over a lake with mountains in the distance. Players had a first-person perspective that provided them with an overview of the virtual environment, and they could look around and see their opponent's scull in the lane that corresponded to their own. In addition to the Unity app, a Voice over Internet Protocol (VoIP) call was established to allow participants to communicate.

The application's third configuration option allows users to compete against an artificial opponent, resulting in the creation of an artificial multiplayer race. The speed of the artificially created opponent was computed in such a way that every 1.25 seconds, the opponent would either speed up or slow down by a random factor

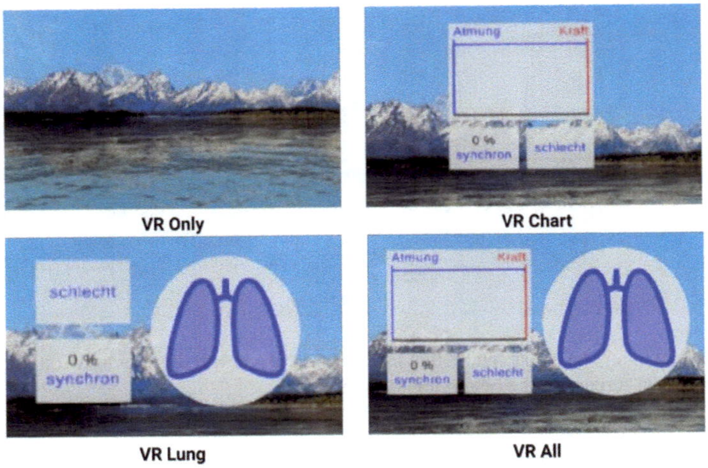

Fig. 3.5 Overview of different visualizations of constant breathing feedback in the application

ranging from 0.15 to 0.3, depending on whether it was in front of or behind the test subject. This was done to simulate the dynamics of the game.

Finally, a fourth scene was built to incorporate breathing feedback with the rowing. Instructions for constant breathing feedback were displayed using three different types of visualizations, where at a time, only one visualization could be activated. Visualization named *VR Chart* displayed continuous breathing biofeedback in VR as a line chart together with the synchronization display. Another possibility for visualization was called *VR Lung* because constant breathing biofeedback was illustrated as lung animation in addition to synchronization display in VR. Finally, a third visualization has at the same time displayed lungs and a chart of constant breathing biofeedback with the synchronization display (Fig. 3.5).

3.2.2 Reading Application

In order to determine which font settings are preferred for reading in VR, the goal of this application was to enable participants to choose both the best and the worst possible combination of the distance between the text and the background, as well as the size and contrast of the text. When participants picked settings for the program, a log file was automatically updated with all of the values they chose.

The Unity application was designed to function as a virtual environment with a simple and empty setting, making it possible to concentrate on reading text. The text distance, font size, and contrast of the text and background could each be adjusted via one of three distinct steps in the virtual world, as illustrated in Fig. 3.6. Just colored objects and sounds were used to guide participants through the various phases so that the virtual environment would not include any additional

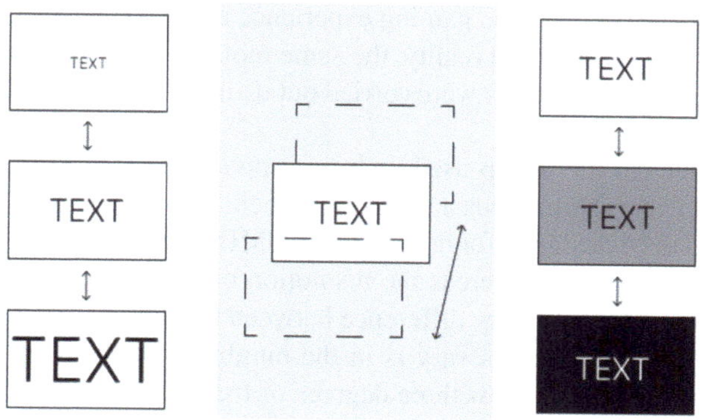

Fig. 3.6 Possible text parameter manipulations: text size, distance, and contrast between text and background

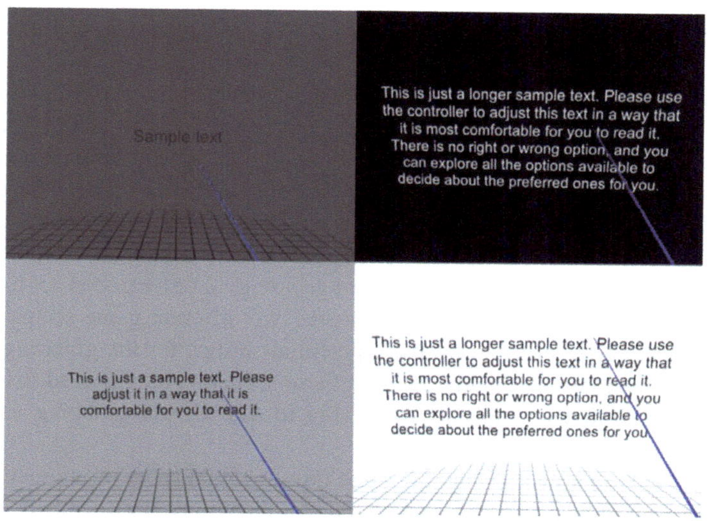

Fig. 3.7 Examples of UI inside a VR application where the user may modify text parameters (text size, distance, and color contrast) for three distinct text lengths using controllers

text that may potentially affect them. Next, there were three distinct lengths of text that were displayed, each of which was determined by the number of words: short (two words), medium (twenty-one words), and long (fifty-one words). Arial was the typeface that was used to display the text, and the spacing and weight were both set to their default values. Examples of virtual scenes are shown in Fig. 3.7. In general, users had the ability to adjust the text settings inside the software in order to set the distance to the text (between 0 and 10,000 mm), the font size (between 5 and 40 pt), and the contrast ratio (from 1:1 to 21:1). Even though each person could only indicate one value at each phase, it was still possible to switch between stages and travel back and forth between stages in order to collect the results that were needed.

In addition, in order to make the gaming experience easier to understand for players who are unfamiliar with virtual reality, the same motions were used in each level of the game, and all manipulations were carried out using controllers that were directly connected with HMDs.

When it comes to the set-up used with this application, due to the fact that the Oculus Go and the Oculus Quest each have their own unique set of specs, the application had to be designed for both of these HMDs. The resolution of the Oculus Go is 1280 by 1440 LCD, whereas the resolution of the Oculus Quest is 1440 by 1600 OLED. This is the primary difference between the two. Another way in which the two virtual reality headsets vary is in the number of degrees of freedom they provide. The Oculus Go has just three degrees of freedom, while the Oculus Quest has six degrees of freedom.

3.2.3 Haptic Application

When considering situations that could benefit from the use of serious games, it is useful to remember that hand motions are frequently incorporated in real-world scenarios. Working in an office, learning, or undergoing rehabilitation are examples of these situations. In some cases, such as sitting at a desk in an office, it is necessary to consider the constraints imposed by the working environment, which limit the types of bodily motions that are acceptable [217]. As a result, the emphasis is on capturing hand gestures, such as typing and gripping, in order to control the device Oculus Quest. During the course of this work, two different game scenarios were developed in Unity in order to examine potential set-ups for the grabbing activity and the typing activity separately. Both of these scenarios emphasized using one's hands to play, with the emphasis on having to move from grabbing to typing, depending on the nature of the game.

In addition, both scenarios had the same four potential types of interaction, with the only two differences being visualization and the choice between hand tracking and controller-based interactions. Each individual user had the opportunity to experiment with all four possible combinations of the following: (1) using controllers while also having their hands visualized; (2) using controllers but without having their hands visualized; (3) using controllers but without any visualization of the controllers, only their hands; and (4) using hand tracking while having their hands visualized.

The first scenario involved playing a game with the objective of sorting five red balls and five blue balls into the appropriate boxes by grabbing the balls and placing them in the appropriate compartments. The second scenario was based on a game in which the player had to retype provided number sequences on a virtual numerical keyboard. They had to do this for a total of five distinct number sequences. Figure 3.8 shows the virtual environment for both scenes (typing and grabbing), as well as graphic representations of all interaction types (with controllers and hand tracking).

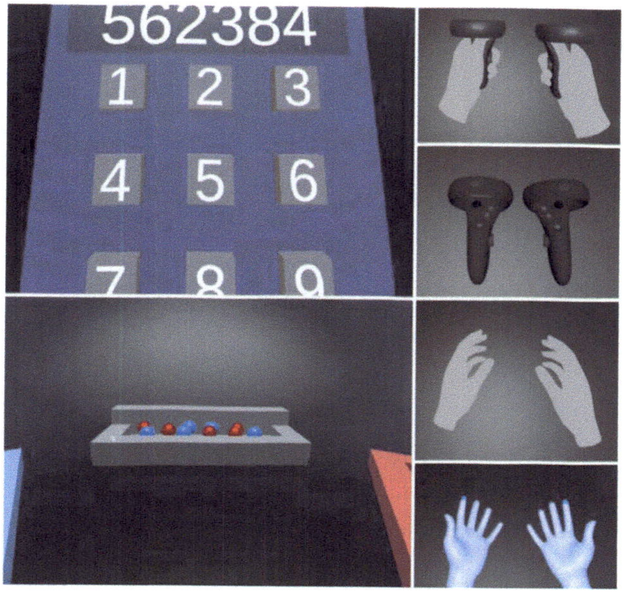

Fig. 3.8 On the left-top is a preview of the typing task, and on the bottom-left is a preview of the grabbing task. On the right, from top to bottom, is an interaction with the visualization of both controllers and hands, visualization of only controllers, visualization of only hands, and interaction with hand tracking

This application was designed to work exclusively with Oculus Quest as an HMD because of the fact that it supports interactions using hand tracking in addition to interactions using controllers.

3.2.4 Data Visualization Application

VR is a medium that offers several opportunities for the display of data, where users could be able to interpret the data better and identify patterns. This comes as a result of the data presentation, which simplifies it so that even those without a background in data may comprehend it. There does not seem to be a significant difference in the total amount of task workload required between conventional visualizations and visualizations in virtual reality, according to the findings of prior research [218]. As a result, the creation of data visualizations for typical use cases such as personal finances might be beneficial to consumers as well. Keeping this objective in mind, an application for data visualization, along with an example of a personal finances overview, was developed.

The Oculus Go HMD was used to display 3D model charts of chosen personal financial statement data based on the 2D personal financial statement used as

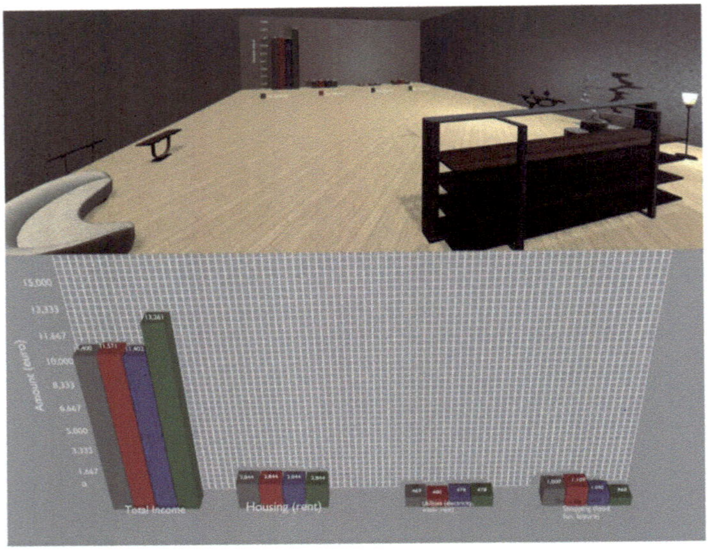

Fig. 3.9 Preview of the virtual environment in VR (up) with visualization of the graph for the Total Expenses over year quarters (below)

prototype data for the Data Visualization Application. This Unity application includes 3D model charts of chosen personal financial statement data, which are shown inside a living space that is completely equipped. The data that were incorporated into 3D and portrayed in VR were column chart representations of Total Income vs. Total Expenses and Total Income versus Total Expenses for each of the year's four quarters, as shown in Fig. 3.9. Pie charts were used to illustrate the Total Annual Income (for each of the four quarters) as well as the Annual Total Expenses (for the first three months of the year).

3.2.5 Social Outside Application

Despite the fact that an increasing number of people have begun to adopt VR, the use of such technology in situations other than private spaces is still uncommon. Nonetheless, because VR devices are now all-in-one solutions that do not require any additional equipment, this technology has the potential to be used in a variety of contexts. There are some examples of VR solutions being used in public settings, such as museums [219] or applications for tourism [220]. However, the presence of spectators who are unaware of what is going on has the potential to degrade the user experience and make people question whether or not this technology is socially acceptable. This Unity application was created to investigate such situations of employing serious games in various social contexts. The application includes a

(a) One Distant Person

(b) Few Distant Persons

(c) Few Close Persons

(d) Many Close Persons

Fig. 3.10 Examples of SSE-VR for 360° videos made available on Youtube by Sygic Travel [222], including a different number of people or level of proximity: (**a**) One Distant Person—Dubai, Safa Park, (**b**) Few Distant Persons—Milan, Arch of Peace, (**c**) Few Close Persons—Paris, Arch of Defense, (**d**) Many Close Persons—Vienna, Christmas Markets

number of Simulated Social Environment (SSE) scenarios in which 360° Videos from YouTube [221] that fulfill certain design criteria were selected to replicate a social environment in VR. All of the selected videos had to be shot from a static point of view with an open perspective, with a diverse number of people visible to the camera at varying degrees of closeness. Figure 3.10 shows the overview of locations selected from the videos providing different settings: one distant person, a few distant persons, a few close persons, and many close persons.

There had to be a way to transition from the SSE-VR world to the VR game environment. The intended transition technique required a transition object, which was a 3D representation of a headset that appeared in the SSE-VR a few seconds after the 360° video began. The user could digitally take it with the controllers, move it to its head, and then release the button to put it on. Scenes shifted, and a loading window appeared to indicate that the transition was taking place. Following the transition, the user could have been in either a static or dynamic game. The static game (shown in Fig. 3.11a) is designed to be played with little interaction, simulating interactions in the office. It just needs one hand gesture to use a controller to point for selecting numbers from 1 to 50 in the correct order as fast as possible. The dynamic game aims to be playable and interactive, like those used in rehabilitation. This game requires the user to perform four different sorts of tasks, as shown in Fig. 3.11b, c.

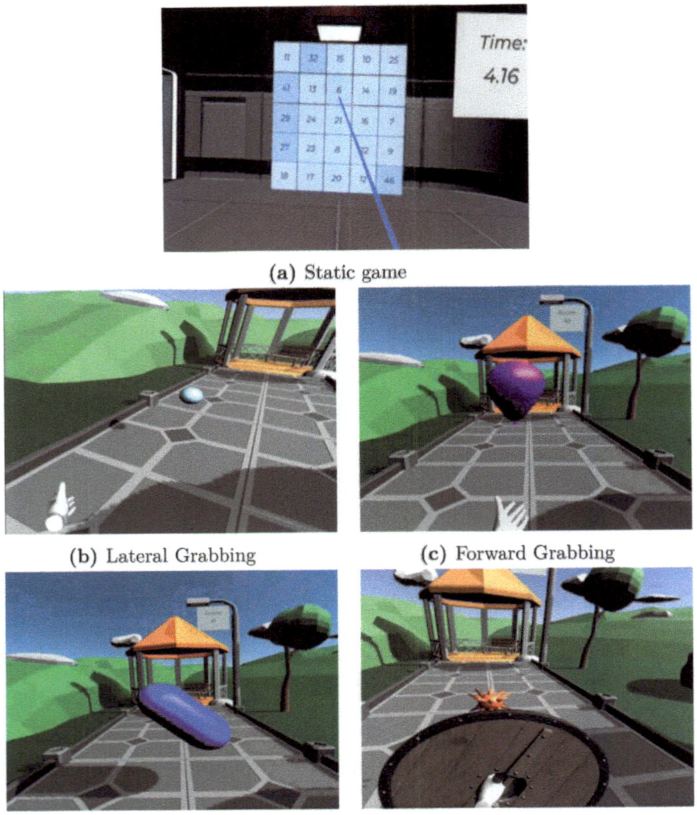

(a) Static game

(b) Lateral Grabbing

(c) Forward Grabbing

(d) Dodging

(e) Shielding

Fig. 3.11 Examples of: (**a**) Static game UI, which is composed of 25 cells containing numbers. Dynamic game UI, examples of the four different kinds of tasks: (**b**) Lateral Grabbing, (**c**) Forward Grabbing, (**d**) Dodging, (**e**) Shielding

3.3 User Studies

The section that follows is an outline of the user studies that were carried out with the aim to provide answers to the research questions that were posed. In the overview, the methodology is summarized, including the specific set-up that was used in each user study. This was done in accordance with the conditions required to answer the research questions and the parameters that were measured. Further, for each user study, it will be described where the experiment was conducted as well as who were the participants in the experiment. However, the ordinal number of user studies is not connected with the ordinal number of the research question, but Fig. 3.1 provides an overview of the connection between user studies and research questions.

3.3.1 Study 1: Influence of Virtual Environments and Conversations on User Engagement During Multiplayer Exergames

The following study aimed to investigate how the VR set-up affects user motivation and engagement (RQ1.1) when compared to the set-up without any additional visual environment. The VR set-up for this study included integrating a virtual environment in VR and the ability for people to converse while playing. The serious game chosen for this study was the Sport Application, as motivation and engagement are important factors in sports. Exergames use gaming motivation to create a more enjoyable experience of physical activity, which can often be exhausting [89]. Additionally, the single play was a very different experience than competing against another player [201], as when players played games with other people, they felt more satisfied. Also, a conversation between players can facilitate the transfer of more information, which can lead to an increase in trust among players [202]. However, it was to be investigated how VR affects players' motivation and engagement in multiplayer exergames, especially when players can communicate with one another. A summary of all details of Study 1 is provided in Table 3.1.

Table 3.1 A summary of the methodology used in User Study 1

	Study 1
Set-up	
Hardware	HTC Vive with PC, a wireless in-ear headset, a large TV and a Augletics Eight ergometer
Software	Rowing VR App
Conditions	
	VR and DTC: with the VE in VR and conversation possibility,
	VR without DTC: with the VE in VR but without the possibility of conversation with another participant,
	No VR and with DTC: with the TV and the possibility of conversation with another participant,
	No VR and no DTC: with the TV and without the possibility of conversation with another participant.
Task	Rowing against the other player in live race.
Parameters	
	Demographics
	Flow (S FSS-2)
	Presence (IPQ)
	Social presence (NMQ)
	Preferred setup
Participants	
Number	36 participants in lab study
Demographics	20 female and 16 male. The average age of the participants was 28.16 years (SD = 6.14, min = 18, max = 47).

3.3.1.1 Procedure: Study 1

The experiment was done in two lab rooms, and during the experiment, each participant was placed in a different room. After being introduced to the experiment, participants filled out a pre-questionnaire regarding demographics including questions about sports and technology experience. Before each race, there was a one-minute exploration time where participants could familiarize themselves with equipment and the atmosphere. Participants could use this time to practice a few strokes and, depending on the condition, explore the virtual environment in VR, or watch the television (TV) show progress bars and understand their feedback mechanisms. Participants were instructed to race against each other four times throughout the experiment, with having 15-minute breaks between races to recover and reflect. Each race was 300 meters, which is the minimal distance adequate to experience racing. The sessions were short enough to prevent fatigue of participants but long enough to simulate the race.

3.3.1.2 Conditions: Study 1

Four distinct conditions were established with the intention of exploring the impact of virtual environments in VR compared to typical gym sessions with only a large TV for displaying a progress bar of each player in meters used for the condition without VR, and the possibility of a dyadic telephone conversation (DTC) between players in multiplayer exergames:

- *VR and DTC*: with the visual environment in VR and conversation possibility
- *VR without DTC*: with the visual environment in VR but without the possibility of conversation with another participant
- *No VR and with DTC*: without a visual representation of the virtual environment but with the TV and the possibility of conversation with another participant
- *No VR and no DTC*: without a visual representation of the virtual environment but with the TV and without the possibility of conversation with another participant

3.3.1.3 Parameters: Study 1

The use of subjective methods was driven by the self-determination theory, flow theory, and the experience of presence. The relationship between user experience and theories was further explained in Sect. 2.2. Therefore, for the measurement of flow with respect to physical activities the SHORT Flow State Scale (S FSS-2) [223] was used, while for measurement of presence in VR, the Igroup Presence Questionnaire (IPQ) [224] was used. As in this experiment, the emphasis is placed on multiplayer, additionally to the presence, the feeling of the social presence was measured with the Networked Minds Measure of Social Presence (NMQ) [225]. At the end of the experiment, participants were asked to rate what option (with or without VR and DTC) they preferred the most.

3.3.1.4 Participants: Study 1

Overall, 36 people (20 female and 16 male) participated in the study, with the average age of 28.16 years (SD = 6.14, min = 18, max = 47). Regarding their sports activity they have reported doing 3.47 hours of exercise per week on average (SD = 2.88), while regarding their previous experience with VR technical systems, the majority (69,4 %) reported having little to no experience with VR.

3.3.2 Study 2: Influence of UI Complexity and Positioning on User Experience During VR Exergames

As video games and VR are comparable and share some of the research, the UI research that has been done on video games might potentially be transferred to VR [197]. On the other hand, when compared to the implementation and the way information is represented in the non-gaming environment, there are some significant differences. So far, recommendations have been made for VR interfaces in general, but there has not been much research done on UI in VR serious games. Even though a significant amount of research has been conducted in the fields of gaming, VR, and UI, very little attention has been paid to a combination of these topics. A summary of all details of Study 2 is shown in Table 3.2.

Serious games with active users, such as exergaming, are particularly challenging because of health and safety concerns [68], as well as the fact that sensory feedback from the game must be immediate. As a result, for the research on how to show parameters in UI for VR, the Sport Application was chosen as an example of an exergame, with different settings for parameter visualization and the goal of exploring the influence on user experience (RQ2.2).

3.3.2.1 Procedure: Study 2

The experiment was conducted in an experimental room of TU Berlin, where participants were invited by a moderator. Participants were given some time to get used to the ergometer as a device without a VR after being briefed on the purpose of the experiment. During that time, they were also tasked with choosing a comfortable pace in kilometers per hour that they could maintain throughout all four exercises. In addition, participants were given time at the start of each of the four sessions prior to the exercise to explore the virtual environment and become acquainted with the UI. Following the exploratory phase, each participant was given a pre-questionnaire. After that, it was time for the participants to begin their exercise, with the goal of maintaining the same pace throughout. When working out on the ergometer, a common training goal is to maintain a consistent pace throughout the activity. Despite the fact that it was relatively short in terms of rowing in general, the

Table 3.2 A summary of the methodology used in User Study 2

	Study 2
Set-up	
Hardware	HTC Vive with PC and a Augletics Eight ergometer
Software	Rowing VR App
Conditions	
	Gamified cockpit: at the front of the rowing scull with gamified visualization of metrics,
	Gamified coach boat: that follows the player with gamified visualization of metrics on a screen in the VR,
	Digitalized cockpit: at the front of the rowing scull with digital visualization of metrics,
	Digitalized coach boat: that follows the player with digital visualization of metrics on a screen in the VR.
Task	Task to choose a comfortable speed in kilometers per hour, which they would be able to maintain throughout all four workouts.
Parameters	
	Demographics
	General motivation for sports (Mpam-r)
	Flow (S FSS-2)
	User Experience (UEQ-S)
	Readability and clearness of UI (7-point Lickert scale)
	Support to the user during the workout (7-point Lickert scale)
	Users personal impression of success (7-point Lickert scale)
	Preferred setup
Participants	
Number	26 participants in lab study
Demographics	16 male and 10 female. The average age of the participants was 28.54 years (SD = 7.38 , min = 18, max = 54).

exercise distance of 300 meters was chosen in order for them to feel immersed in the experience but not to exhaust them. The participants were also given a 10-minute break between sessions to relax and reflect on their overall experience.

3.3.2.2 Conditions: Study 2

Four conditions were defined with the goal of investigating the impact of UI positioning and complexity on user experience and performance, as visualizations of:

- *Gamified cockpit*: as a cockpit at the front of the rowing scull with gamified visualization of metrics
- *Gamified coach boat*: as a coach boat that follows the player with gamified visualization of metrics on a screen in the VR
- *Digitalized cockpit*: as a cockpit at the front of the rowing scull with digital visualization of metrics

- *Digitalized coach boat*: as a coach boat that follows the player with digital visualization of metrics on a screen in the VR

3.3.2.3 Parameters: Study 2

The pre-questionnaire requested demographic information, frequency of participation in sports, and amount of prior familiarity with such technology. In addition, the MPAM-R questionnaire [135] was used to assess overall motivation for sports in five categories: appearance, competence, interest, sociability, and fitness. Furthermore, the method included aspects of flow theory and user experience, so the Short Flow State Scale was used (S FSS-2) [226, 227], as well as the short User Experience Questionnaire (UEQ-S) [124], to evaluate user experience in both pragmatic and hedonic qualities.

A few customized questions with a 7-point Likert scale were also included to evaluate the following parameters: readability and clarity of the UI, support to the user throughout the exercise, and the user's own sense of success in accomplishing the objective. Following the experiment, participants were asked to choose their preferred set-up, propose metrics for the UI, and rate their overall performance. Finally, to assess performance, the current speed was captured every 2 seconds throughout the exercise to compute a mean speed deviation relative to the goal speed.

3.3.2.4 Participants: Study 2

Overall data of 26 people (16 male and 10 female) who participated in the study were analyzed. They have reported being, on average 28.54 years old (SD = 7.38, min = 18, max = 54), with practicing some exercise 4.94 hours per week (SD = 1.99). Furthermore, the majority of participants reported having little to no prior experience with rowing as a physical exercise (65.4%) and also had little to no prior familiarity with VR (69.2%).

3.3.3 Study 3: Influence of Network Delay in Virtual Reality Multiplayer Exergames

Multiplayer virtual reality (VR) exergames are technically feasible due to the fact that the parameters determining interactions between players are sent over a network and shown inside the player's vision of the virtual environment. As a result, the speed at which those parameters are exchanged is critical, and it is essential to ensure that an undesirable delay will not influence the user experience. When it comes to controlling characters, the research suggests that values of fewer than

Table 3.3 A summary of the methodology used in User Study 3

	Study 3
Set-up	
Hardware	HTC Vive with PC and a Concept2 Model D rowing ergometer
Software	Rowing VR App
Conditions	
	Own L and Opp. 's L: as player's low (30ms) and opponent's low (30ms)
	Own L and Opp. 's M: as player's low (30ms) and opponent's medium (100ms)
	Own L and Opp. 's H: as player's low (30ms) and opponent's high (500ms)
	Own M and Opp. 's L: as player's medium (100ms) and opponent's low (30ms)
	Own M and Opp. 's M: as player's medium (100ms) and opponent's medium (100ms)
	Own H and Opp. 's L: as player's high (500ms) and opponent's low (30ms)
	Own H and Opp. 's H: as player's high (500ms) and opponent's high (500ms)
Task	Rowing agains the artifical opponent with different delay settings
Parameters	
	Demographics
	QoE for the delay of oneself and of the opponent (10-point Likert scale)
	Presence (IPQ)
	Flow (S FSS-2)
Participants	
Number	23 participants in lab study
Demographics	17 male and 6 female. The average recorded age was 25.57 years (SD = 3.41 years).

150 milliseconds are optimal [159]. However, a reduction in the sensation of being present has been shown to occur in virtual settings at values of delay between 50 and 90 milliseconds [160]. Games that are particularly sensitive to delay are those in which different perceptions of delay can lead to motion sickness or a feeling of different results, such as racing. Therefore, the application used for this study is the Sport Application, as an exergame where race is a part of the scenario. With this user study (shown in Table 3.3), the aim is to investigate the effects of delays in the player's network and the opponent's network, as well as the user's perception of delay (RQ1.2) and its effect on the overall experience.

3.3.3.1 Procedure: Study 3

The participants in the experiment were briefed on the equipment and the methodology that would be used at the start of each session of the experiment. Participants in the test were given the challenge of rowing against a computer-generated opponent, while the delay between their strokes was adjusted. The introduction of delays only had an effect on receiving rowing parameters (speed and distance), but the rendering and visualization of the virtual world were done locally and were determined by the location of the HMD. The participant's task was to row seven times over sessions and evaluate the perceived delay. The test participant was given questionnaires to fill out after each of the seven sessions of rowing that were conducted.

3.3.3.2 Conditions: Study 3

The research included the use of three different amounts of delay: 30, 100, and 500 milliseconds. These three different amounts of delay will be referred to in the following as LOW, MEDIUM, and HIGH, respectively. In all, there were seven circumstances that were specified for the experiment, each of which included a unique combination of own-player delay and opponent delay. The situations in which the player's delay was MEDIUM, and the opponent's delay was HIGH, as well as the conditions in which the player's delay was HIGH and the opponent's delay MEDIUM, were not employed in order to prevent extreme tiredness brought on by the physically demanding activity.

All delays used for the user study are:

- *Own L and Opp. 's L*: as player's low (30 ms) and opponent's low (30 ms)
- *Own L and Opp. 's M*: as player's low (30 ms) and opponent's medium (100 ms)
- *Own L and Opp. 's H*: as player's low (30 ms) and opponent's high (500 ms)
- *Own M and Opp. 's L*: as player's medium (100 ms) and opponent's low (30 ms)
- *Own M and Opp. 's M*: as player's medium (100 ms) and opponent's medium (100 ms)
- *Own H and Opp. 's L*: as player's high (500 ms) and opponent's low (30 ms)
- *Own H and Opp. 's H*: as player's high (500 ms) and opponent's high (500 ms)

3.3.3.3 Parameters: Study 3

Before participating in any of the rowing sessions, the participants were asked to fill out a demographics questionnaire, which included questions about their age, gender, and the frequency with which they play multiplayer video games on a scale of one to five. The participants in the experiment were then given questionnaires to fill out after each of the seven sessions of rowing. This experiment's initial questionnaire was designed only for the purpose of gathering information about participants' QoE. On a Likert scale with ten points, the participant was asked to rate how much delay they and their opponent experienced during a certain session. Additionally, the Igroup Presence Questionnaire (IPQ) was used to measure the sense of presence in VR [224], while for measuring flow occurring during physical exercise, the 5-point FSS-2 scale was used [227].

Table 3.4 A summary of the methodology used in User Study 4

	Study 4
Set-up	
Hardware	HTC Vive with PC, a Augletics Eight ergometer and a BreathZpot sensor
Software	Rowing VR App
Conditions	
	No VR: an option that is not using VR and without breathing biofeedback, VR Only: an option in VR but without breathing biofeedback, VR Chart: constant breathing biofeedback as line chart and synchronization display in VR, VR Lung: constant breathing biofeedback as lung animation and synchronization display in VR, VR All: all visualizations of constant breathing biofeedback displayed in VR.
Task	Maintain correct breathing rythem to breath out when the pull starts.
Parameters	
	Demographics
	Flow (S FSS-2)
	Presence (IPQ)
	User Experience (AttrakDiff)
	Ranking different vizualizations by sympathy and helpfullness
Participants	
Number	23 participants in lab study
Demographics	10 female and 13 male. The mean age was 23.67 years (SD = 4.887, min age = 17, max = 36).

3.3.3.4 Participants: Study 3

The research was carried out on a total of 23 participants, with 17 men and 6 female participating as test subjects. The average age reported was 25.57 years, while the standard deviation of ages was 3.41 years.

3.3.4 Study 4: Influence of Constant Visual Biofeedback on User Experience in Virtual Reality Exergames

Feedback has a strong impact on learning and achievement, and because of that, it is often used in sports [228]. Biofeedback uses feedback to control biological functions in an indirect manner; for example, biofeedback signals include brainwaves, heart rate, pain perception, and breathing (respiration). In general, there are applications for respiratory biofeedback [229], as it is useful for training and controlling breathing rate. Therefore, the Sport Application was used in this user study (as shown in Table 3.4). The purpose of this research was to explore the motivation for the usage of biofeedback in VR set-up (RQ 1.1) and to see how various types of continuous visual biofeedback affect the flow and capacity to maintain a given breathing rhythm (RQ 2.1).

3.3.4.1 Procedure: Study 4

The target audience included beginning rowers as well as those rowers who were unfamiliar with the breathing visualization method. The participants were chosen in this manner because the likelihood of progress in terms of breathing pattern was projected to be greater for this group compared to rowers who previously mastered this pattern [230].

Participants completed a demographics questionnaire and learned the breathing task before rowing; it stated that participants should exhale as the pull starts. During a rowing exercise, correct breathing was calculated as a percentage of total breaths. Rowing sessions begin with three-second countdowns, and the race ends after 300 meters.

3.3.4.2 Conditions: Study 4

During the course of the experiment, each participant performed five sessions of rowing, during which one of the following experimental conditions was presented in each session:

- *No VR*: as an option that is not using VR and without breathing biofeedback
- *VR Only*: as an option in VR but without breathing biofeedback
- *VR Chart*: as constant breathing biofeedback as line chart and synchronization display in VR
- *VR Lung*: as constant breathing biofeedback as lung animation and synchronization display in VR
- *VR All*: as all visualizations of constant breathing biofeedback displayed in VR

To determine whether or not the VR experience had any impact on its own, a baseline condition was used that did not include VR or breathing biofeedback. Rest periods were used in between rowing sessions to avoid overexertion, and after each rowing session, participants answered questionnaires.

3.3.4.3 Parameters: Study 4

In order to measure the flow that occurred during the various rowing sessions (conditions), the SHORT Flow State Scale (S FSS-2) [227] was utilized. Additionally, the participants were asked to evaluate their sense of presence using the general item G1 from the "Igroup Presence Questionnaire" (IPQ) [224]. Furthermore, the short 10-item version of the AttrakDiff was used in order to record the user experience while rowing in VR [123]. Additionally, two particular questions were designed to rank conditions based on preference (represented as the following question: "Which interface did you like the most? Please rank the interfaces from highest to lowest.") and helpfulness (represented as the following question: "Which interface was the most helpful one for keeping the recommended breathing rhythm? Please rank the interfaces from highest to lowest.").

3.3.4.4 Participants: Study 4

There were a total of 23 participants (10 females and 13 males). The mean age was 23.67 years (SD = 4.887, min = 17, max = 36). Only three of the participants had some rowing expertise, while the rest were novices.

3.3.5 *Study 5: Influence of Reading in Virtual Reality on User Experience—Finding Values for Text Distance, Size, and Contrast*

Reading is an important aspect of most jobs; thus, it seems meaningful to incorporate some of that reading into work settings that are simulated in VR [231]. There are a lot of different ways that text might be used inside the world of VR. Therefore, this user study was done in order to discover what complexity of text characteristics is necessary for comfortable reading and positioning of text in VR (RQ 2.2). As part of this study, participants were given the opportunity to customize their preferred settings so that the results might be more accurate. The size of the text, the distance between the user and the text, and the color contrast of the text and the background were the primary focuses of the research. Even while studies have previously been conducted in the field of readability on the Web [232], not much attention has been paid to the question of how such standards translate to text included inside VR.

3.3.5.1 Procedure: Study 5

To ensure that each participant had their own time to dedicate to the experiment, they were offered the option of signing up for a specified time slot in the university laboratory. At the start of the study, participants signed a consent form and were provided with a pre-questionnaire about demographics.

Furthermore, participants were then given an overview of two Oculus (Go and Quest) devices with different resolutions, controls, and interactions that were necessary for the experiment. Each participant got the chance to experience both devices using a training program with identical interactions, and for it, no data were collected to allow the participant to examine and experiment with all of the various ways in which the controls could be applied and interacted with. Furthermore, participants were given the option of adjusting the HMD settings to the degree of comfort that was best for them at the time. Because Oculus allows for it, those who wear glasses may do so while using the HMD.

Table 3.5 A summary of the methodology used in User Study 5

	Study 5
Set-up	
Hardware	Oculus Go, Oculus Quest
Software	Readability VR App
Conditions	
	Go and Short: showing the short text of 2 words, Go and Medium: showing the medium text of 21 words, Go and Long: showing the long text of 51 words, Quest and Short: showing short text of 2 words, Quest and Medium: showing medium text of 21 words, Quest and Long: showing long text of 51 words.
Task	For each condition: set text parameters to get the best and the worst possible readability
Parameters	
	Angular size (dmm) Contrast ratio Afinity for Technology (ATI) Emotional assesment (SAM) System Usability (SUS)
Participants	
Number	22 participants in lab study
Demographics	12 male and 10 female. The average age of the participants was 28.41 years (SD = 9.56 , min = 19, max = 62).

Finally, after the training, participants were asked to read different texts and to set the parameters of distance and contrast for the best and worst possible readability. In those settings, participants were asked to evaluate the user experience as well as to rate the overall usability score of the system, as it is shown in summary Table 3.5.

3.3.5.2 Conditions: Study 5

There were a total of six conditions, each of which was a unique combination of two Oculus devices and one of the three possible text lengths:

- *Go and Short*: as Oculus Go showing the short text of two words
- *Go and Medium*: as Oculus Go showing the medium text of 21 words
- *Go and Long*: as Oculus Go showing the long text of 51 words
- *Quest and Short*: as Oculus Quest showing short text of 2 words
- *Quest and Medium*: as Oculus Quest showing medium text of 21 words
- *Quest and Long*: as Oculus Quest showing long text of 51 words

Participants were given two tasks to complete: The first task required them to configure the text in such a way that it had the highest possible readability, and the second task required them to configure the text in such a way that it had the lowest possible readability. Participants were required to set up the text settings for each condition, which included the distance, font size, and contrast of the text.

3.3.5.3 Parameters: Study 5

Several questionnaires were included throughout the study. In addition to demographics, the pre-questionnaire included the Affinity for Technology Interaction (ATI) [233] questionnaire to determine users' tendency to participate in technology interaction actively. Also, participants were required to evaluate the Self-Evaluation Manikin (SAM) [121] as an emotion assessment instrument for enjoyment, arousal, and dominance after each exercise. Furthermore, after configuring text parameters for all conditions, participants were asked to rate the usability of reading in VR using the System Usability Scale (SUS) [119] and to evaluate general readability in VR, with the option to make any extra remarks.

3.3.5.4 Participants: Study 5

The research included a total of 22 participants (12 males and 10 females). The participants' average age was 28.41 years (SD = 9.56, min = 19, max = 62). The majority of participants (86.4%) had some familiarity with virtual settings, whereas just three (16.6%) had never tried VR before, and no one was an expert. Furthermore, participants claimed that the majority of them read between 2 and 4 hours each day on average (77.3%).

3.3.6 *Study 6: Influence of Hand Tracking as a Way of Interaction in Virtual Reality on User Experience*

Controllers are gradually being eased out in favor of new HMD features that allow users to interact with items by monitoring their hands. This new possibility has the potential to contribute to many interesting use cases in virtual reality in the gaming sector, as well as in other industries where virtual reality is utilized for serious games, such as learning [73], medical [74], or manufacturing [75]. Hand gestures play an important role in these disciplines, which have traditionally been associated with precision. However, in order to get accurate simulations and results, it is also essential to guarantee a high level of quality and provide a satisfying experience for users.

In order to explore the possibilities of hand tracking itself, the application used for this experiment was the Haptic Application as it replicated typing and grabbing as two interactions, often used in office environments. Therefore, this study aims to investigate how different interaction types, such as hand tracking or interaction via controllers (RQ 2.1), depending on different task types, such as grabbing or typing (RQ 2.2) in VR, influence user experience. Summary of the study is given in Table 3.6.

Table 3.6 A summary of the methodology used in User Study 6

	Study 6
Set-up	
Hardware	Oculus Quest
Software	Haptics App
Conditions	
	C_H: visualized controllers and hands, C_noH: visualized controllers, but no visualization of hands, C_OnlyH: visualized hands, but no visualization of controllers, HTrack: visualized hands while using hand tracking.
Task	Sorting objects by color and retyping number sequence.
Parameters	
	Demographics Affinity for technology (ATI) Emotional assesment (SAM) Presence (IPQ) System Usability (SUS)
Participants	
Number	20 participants in lab study
Demographics	8 female and 12 male, with the average age of the participants was 30.20 years (SD = 12.74, min = 19, max = 62).

3.3.6.1 Procedure: Study 6

Participants were greeted at the start of the experiment, and as part of the introduction, it was said that the assignment was to play two virtual reality games, each with a separate task, using controllers and hand tracking as methods of interaction. Before experiencing any VR condition, participants had to fill out a consent form and a questionnaire about their demographics, previous experience with virtual reality, and hand tracking skills.

As part of the introduction to VR and interactions, participants were given the opportunity to become acquainted with the virtual environments of both games. This was accomplished by teaching them how to interact with controllers and hand tracking and giving them the opportunity to do so. The moderator gave permission to begin the experiment when the participants demonstrated their confidence in their abilities to perform the learned interactions. Each participant took part in a total of eight different conditions because they played two games with different task types, grabbing or typing four times with each of the different interaction types. The moderator asked the participants about the visualizations they had seen and the behaviors they demonstrated throughout the session at the start of each condition. Participants were given questionnaires to fill out after each condition, and at the end of the experiment, they were asked to rate and rank all possible interactions.

3.3.6.2 Conditions: Study 6

In the application, both games, depending on the sort of game to grasp or type, placed a great focus on being played with hands, and both games featured the same four possible types of interactions:

- *C_H*: as visualized controllers and hands
- *C_noH*: as visualized controllers, but no visualization of hands
- *C_OnlyH*: as visualized hands, but no visualization of controllers
- *HTrack*: as visualized hands while using hand tracking

3.3.6.3 Parameters: Study 6

In the beginning, questions regarding the participants' demographics were asked. In addition, questions regarding the participants' tendency to actively engage in interaction with technology were included in the Affinity for Technology Interaction (ATI) Scale questionnaire [233].

To get an accurate evaluation of the participants' feelings, they were given a Self-Assessment Manikin (SAM) to complete at the conclusion of each condition [121]. In addition, after each condition, participants judged their sensations of presence and realism using a standardized measure of a perceived presence called the Igroup Presence Questionnaire (IPQ) [234]. After completing all conditions, participants were asked to rate the usability of reading in VR using the System Usability Scale (SUS) [119] for hand tracking on a separate scale for each of the grasping and typing activities. At the end of the study, participants were given one additional questionnaire to complete, in which they were asked to rank the quality of the four separate interactions they had during the study on a scale of best to worst.

3.3.6.4 Participants: Study 6

There were a total of 20 participants (8 females and 12 males). The average age of the participants was 30.20 years old (SD = 12.74, min = 19, max = 62). Further on, four of the participants had no prior experience with VR technology, while the other participants had some exposure to VR in the past; nevertheless, none claimed to be an expert in the field. Also, nine of the participants claimed to have no prior experience at all with using hand tracking for interaction purposes, while the other participants either had a limited amount of experience or considerable experience, but none reported being an expert with hand tracking.

Table 3.7 A summary of the methodology used in User Study 7

	Study 7
Set-up	
Hardware	Oculus Go vs. printed Financial Report
Software	Finance App
Conditions	
	Personal finance statement as 3D vizualization in VR
	Personal finance statement as 2D vizualization in paperwork
Task	Anwser questions (such as "Find the quarter in which the least amount of expenses was spent.") about the finance situation based on given data.
Parameters	
	Demographics
	Presence (IPQ)
	User Satisfaction (USE)
	System Usability (SUS)
Participants	
Number	23 participants in lab study
Demographics	5 females and 18 males within different age group of range from 18 to 54 years old.

3.3.7 Study 7: Exploring Visualizations for Financial Statements in Virtual Reality

It is hard to see large amounts of data on a 2D screen without missing some information, so data visualization tries to make it simpler by analyzing data and visually expressing information to users. However, immersive technologies, in particular VR technologies, have opened new opportunities with new approaches for data visualization [64]. Since VR has an opportunity to add an additional dimension to the visualization process [65], 3D visualization could be more helpful in enabling the user to better comprehend the data they see compared to conventional 2D visualization. So, the goal of this user study was to look into different ways to visualize information and compare traditional 2D on-paper visualization with 3D in VR.

With this aim, the Data Visualization Application was used for the user study, and the use case was around personal finance, as these scenarios could provide enough opportunities for users who are not experts in the field still to understand the tasks and motivation for the data analysis (shown in Table 3.7). So, the focus of the study was placed on the users and how different users experience 3D financial statements in VR depending on their age, gender (RQ 4.1), and previous experience with VR (RQ 4.2).

3.3.7.1 Procedure: Study 7

The experiment was conducted with participants in the lab, where the equipment had been set up for usage. Following a short introduction and a review of the processes, the participant was asked to sign a permission form. Following that, a demographics questionnaire was gathered to provide a complete profile of the person. Time was recorded as a measure of performance for each condition for both two tasks that were completed by each participant. Participants filled out questionnaires after completing each assignment with the goal of exploring how financial information may be displayed in VR and focusing on user experience. Each participant was given a maximum of 30 minutes for their time slot during the experiment; thus, there was no opportunity to rest between the two exercises.

3.3.7.2 Conditions: Study 7

In both tasks, participants were given assignments in the form of questions to answer using either paper or VR in order to assess and compare the two different methods. Scenarios included the following tasks:

- *Task 1* was to find the quarter in which the least amount of expenses was spent, and the third highest income was earned.
- *Task 2* was to find the quarter in which the highest income was earned, least spent on housing and utilities, but most spent on shopping.

3.3.7.3 Parameters: Study 7

In order to measure user experience and preferences, the Igroup Presence Questionnaire (IPQ) [149] was adapted (dimension G1) and used to measure the extent to which users experienced presence in the VR, together with the aim to assess the degree to which a participant felt immersed. Further on, System Usability Scale (SUS) questionnaire [118] and Usefulness, Satisfaction, and Ease of use (USE) questionnaire [117] were used to evaluate the usability and ease of use for both 2D and 3D conditions.

3.3.7.4 Participants: Study 7

There were a total of 23 participants (5 females and 18 males), with an average age of 29.36 years (SD = 7.52, min = 18, max = 54). The majority of participants (69.5%) stated that they had some prior experience with VR and that they would withdraw money and check their bank or financial statement between one and three times per month (45.5%).

3.3.8 *Study 8: Influence of Interactivity and Social Environments on User Experience and Social Acceptability in Virtual Reality*

It is still unusual to see VR devices used in outside settings, but as solutions have become all-in-one, this technology has the potential to be used everywhere. As a result, it is important to understand the factors that motivate people to use and be open to the inclusion of technology in a given context. As a result, as several recent studies show, there has been an increase in interest in the issue in research. The usage of VR seems to be suitable in situations such as a bedroom, the subway, or a train. When individuals are put in places where they are expected to participate in social contacts, such as living rooms or public cafés, the acceptability decreases, so it seems that the social acceptability of VR from the user's viewpoint depends on the scenario and context [173]. This study provided the framework for future research, but it had the limitations of an online study, such as the fact that participants did not engage with the technology they were assessing, and the validity of findings from the user's viewpoint was not addressed.

The aim of the study is to contribute to the growth of knowledge and understanding of the factors that influence the adoption of virtual reality technology in social circumstances (summary with Table 3.8). By using the application that offers different types and kinds of interactivity, such as the developed Social Outside application, the purpose is to investigate social acceptability in order to improve VR experiences (RQ 3.1) by better understanding where and how this technology can be successfully applied and accepted.

3.3.8.1 Procedure: Study 8

The participants were invited to a lab room to participate in the experiment on their own, and everyone received a unique invitation to a specific time slot to do the study in the lab. The moderator welcomed the participants and gave them a summary of the investigation. After signing a consent form, participants were given a pre-questionnaire that questioned their demographic information as well as their prior experience with VR. The next step was to acquaint the participants with the Oculus Quest VR equipment, the controllers, and the interactions required for the experiment. Following this brief introduction, the participant was given a chance to begin the condition as indicated by the moderator. While the participant was answering the questions from before, the moderator put the next condition on the participant's headset in the order that was chosen at random for that participant. A final qualitative questionnaire had both multiple-choice and open-ended questions that participants were required to answer.

Table 3.8 A summary of the methodology used in User Study 8

	Study 8
Set-up	
Hardware	Oculus Quest
Software	Simulated Social VR app
Conditions	
	One Distant Static: one distant person and static game Few Distant Static: few distant persons and static game Few Close Static: few close persons and static game Many Close Static: many close persons and static game One Distant Dynamic: distant person and dynamic game Few Distant Dynamic: few distant persons, dynamic game Few Close Dynamic: few close persons and dynamic game Many Close Dynamic: many close persons, dynamic game
Task	Select numbers in the correct order from 1 to 50 for static game, and grabbing, dodging and shielding from objects in dynamic game.
Parameters	
	Demographics Affinity for technology (ATI) Presence (IPQ) User Experience (UEQ-S) Emotional assesment (SAM) Social acceptability (SAQ) System Usability (SUS)
Participants	
Number	28 participats in lab study
Demographics	21 male and 7 female. The average age was 24.64 years (SD = 2.6, min = 21, max = 31).

3.3.8.2 Conditions: Study 8

In total, there were eight experimental conditions as combinations of four different Social Environments and two different degrees of interactivity:

- *One Distant Static* as playing a static game in front of one distant person
- *Few Distant Static* as playing a static game in front of a few distant persons
- *Few Close Static* as playing a static game in front of a few close persons
- *Many Close Static* as playing a static game in front of many close persons
- *One Distant Dynamic* as playing a dynamic game in front of one distant person
- *Few Distant Dynamic* as playing a dynamic game in front of a few distant persons
- *Few Close Dynamic* as playing a dynamic game in front of a few close persons
- *Many Close Dynamic* as playing a dynamic game in front of many close persons

3.3.8.3 Parameters: Study 8

In addition to demographic information, participants were asked to complete the Affinity for Technology Interaction (ATI) Scale questionnaire [233] to determine their tendency to participate in technology interaction actively. Following each condition, the participant was asked to complete questionnaires that included the iGroup Presence Questionnaire (IPQ) general item that measures the "sense of being there" [235], Self-Assessment Manikin (SAM) expressing three emotional elements [121], Social Acceptability (SAQ) built on top of the one used for rating social acceptability from the performer's perspective [172], and Short User Experience (UEQ-S) Questionnaire [125].

After playing through all of the different situations and rating them, the participant was then asked to rate the usability of the VR application using the System Usability Scale (SUS) [119]. A final qualitative questionnaire included topics such as the graphical quality of the 360-degree video, appropriateness of social surroundings and VR games, usage of VR in real-world social situations, and potential future uses of VR technology in social settings.

3.3.8.4 Participants: Study 8

A total of 28 people participated (N = 28, 21 males, 7 females), and the average age was 24.64 years (SD = 2.6 years, min. 21 years, max. 31 years). The participants had an Affinity for Technology Interaction that averaged 4.55 (SD = 0.69). Additionally, nine participants had reported having no prior experience with VR, five described themselves as not at all familiar, four as low familiar, 14 as averagely familiar, four as very familiar, and one as extremely familiar with VR technology.

3.4 Data Analysis

For all of the studies, when it comes to data gathering and preparation, the same approach was used, with the specifics of the data results discussed in the next sections. Overall, data collection was conducted in accordance with ethical principles for medical research involving human subjects proposed by the World Medical Association (WMA) declaration of Helsinki. This is the case; as for all studies, people have participated in and tested out VR serious game applications.

The approach used to gather answers/data is a mixed strategy with a greater emphasis on quantitative research rather than qualitative research. The requirement to quantify and understand causal linkages between variables, which is critical to finding appropriately constructed solutions, motivated the decision to do quantitative research [236]. For all investigations, a within-subject design with condition randomization was used due to the number of experimental conditions (between 2 and 8) and a target sample of more than 20 participants. Furthermore, Web-based questionnaires with multiple-choice scales and open-ended questions were used to collect and evaluate the study's results. Scales have been included in the analyses in their original form and were not converted to ensure the usage of the standardized methods, as well as to ensure the possibility of comparing the results within scale dimensions and standards.

All data sets were evaluated for outliers during data preparation; however, because moderation was done straight from the lab room, there were only a few outliers. Still, in Study 3, a total of 27 participants were recruited. Following the initial speed deviation calculation, one person was eliminated from the study because his performance deviated too much from the mean, with highlights for his rowing speed indicating performance values that differed by more than double the standard deviation from the mean value. This suggested that the participant was not following the study's goal of maintaining rowing speed, so his results were completely eliminated.

The statistical program IBM SPSS was used to do descriptive and inferential statistical analysis, and with it all of the quantitative data was then categorized and examined. It was decided that the most suitable statistical test would be an Analysis of Variance (ANOVA), because it has higher statistical power compared to nonparametric tests, and it is also quite robust to violation of normality [237]. It enables the evaluation of the differences in mean values across groups that have been differentiated based on independent variables, hence giving information on their effect on some dependent variables (for example: presence, emotions, usability, social acceptability, and others). Further on, the Friedman's Two-Way Analysis was done where dependent variables were measured in rankings, as it is a nonparametric test for finding differences in treatments across multiple attempts.

3.5 Summary

This chapter provided an overview of the methodologies that were used to provide answers to research questions exploring different influencing factors. To reach this goal, an overview of the VR applications that have been made and user studies that have been conducted to find out what factors affect the VR experience was shown.

Overall, five different applications, each one representing a different scenario of VR serious games, were created. Each of the games was designed with the intention that it could be expanded into the real market as a serious game. In addition, each of the games was constructed in such a way that it offered full control over the code and parameters of the game, which is something that market-ready game solutions do not offer. This work could not have covered all of the possible scenarios because VR serious games could have been found in an endless number of different areas. Instead, it selected those scenarios that could offer a variety of tasks and possible reasons why users might want to use a particular VR serious game. Scenarios that were selected for VR serious games have included some of the physically dynamic activities, such as sports, as well as some of the more static games that replicate opportunities for the use of serious games in office environments, such as reading, typing, or data analysis. In Sect. 3.2, it was provided an in-depth explanation of each application and the set-up solution for it, along with samples of UI visualizations in VR and the hardware that is required for it. These applications were further on used in user studies to provide a testing environment for the research that has been conducted.

Regarding the subjective user testing, all together, eight studies have been carried out, each focusing on different aspects of experience based on different applications and exploring another influencing factor. Section 3.3 provided insights into each of the studies, summarizing the motivation, procedure, conditions, and parameters of the studies, while also reporting on the participants who took part in the tests. For all studies it is the same that they all took place in a laboratory setting, having invited participants with different backgrounds and using questionnaires as a method to measure the effects of particular conditions. Also, all studies took care that conditions were executed in a randomized order for each participant to avoid order effects, and moderators took care that the data collection was conducted in accordance with ethical principles for medical research involving human subjects proposed by the World Medical Association (WMA) declaration of Helsinki.

In the appendix, there is provided a complete overview with Table C.2, which summarizes all of the user studies.

Chapter 4
System IF

4.1 Results

This chapter aims to investigate and comprehend which technological factors and in what way they can influence the user experience. Because of the fact that HMDs are now the most extensively used VR devices on the market and that the research took place in the laboratory, the emphasis of this chapter is centered on two subcategories: device-related system IFs and network-related system IFs.

In particular, the aim is to explore whether VR set-up influences motivation for serious VR gaming (RQ1.1) and, in the case of the multiplayer scenario, what influence the network delay has on the user's perception of experience (RQ1.2). As a result, three user studies were conducted: Study 1, Study 3, and Study 4 (an overview of the studies is provided in Table C.2), the findings of which are presented below.

In order to determine statistically significant differences, an Analysis of Variance (ANOVA) was performed to investigate the effects between conditions. Further on, Friedman's Two-Way Analysis was done where dependent variables were sympathy rank and helpfulness rank. All findings and effects will be reported in the following sections together with the overview shown in Tables 4.1, 4.2 and 4.3 that give an overview of the statistically significant results of the ANOVA and the Friedman tests.

4.1.1 Impact of VR Device

4.1.1.1 Presence

When comparing set-ups, one with using VR and the other without, for multiplayer exergame based on rowing in Study 1, there have been found significant differences in feeling of presence measured by questionnaires S FSS-2, IPQ, and NMQ.

T. Kojić, *User Experience for Serious Games in Virtual Reality*, T-Labs Series in Telecommunication Services, https://doi.org/10.1007/978-3-031-75530-9_4

Table 4.1 Effects of *VR* and *DTC* on S FSS-2 perception of *time*, on IPQ *general presence G1*, on IPQ *spatial presence SP*, on IPQ *involvement INV*, on IPQ *experienced realism REAL*, and on NMQ *perceived comprehension PE-C*

Parameter	Effect	df_n	df_d	F	p	η_G^2
VR	time	1	35	4.93	.033	0.12
VR	G1	1	35	80.53	< .001	0.69
VR	SP	1	35	83.25	< .001	0.70
VR	INV	1	35	89.02	< .001	0.71
VR	REAL	1	35	20.51	< .001	0.37
VR	PE-C	1	35	10.90	.009	0.23

Table 4.2 Effects of all five different conditions on the feeling of flow from FFS-2 (Flow) and percentage of breathing correctly while exercising (Breathing) calculated with ANOVA

Parameter	Effect	df_n	df_d	F	p	η_G^2
VR	Flow	1	22	5.16	0.001	0.19
VR	Breathing	1	22	2.597	0.042	0.11

Table 4.3 Effects of player's network delay (Own Delay) and opponent's network delay (Opp.'s Delay) on IPQ general presence (PRES), IPQ spatial presence (SP), IPQ involvement (INV), and IPQ experienced realism (REAL), on the perception of flow from FFS-2 (flow), on the perception of own delay (Own Delay) and perception of opponent's delay (Opp.'s Delay), and on overall QoE (QoE)

Parameter	Effect	df_n	df_d	F	p	η_G^2
Own Delay	PRES	2	154	8.974	.000	0.104
Own Delay	SP	2	154	3.733	.026	0.046
Own Delay	INV	2	154	5.992	.003	0.072
Own Delay	REAL	2	154	7.208	.001	0.086
Own Delay	Flow	2	154	7.132	.001	0.085
Own Delay	Own Delay	2	154	45.515	.000	0.372
Own Delay	Opp.'s Delay	2	154	10.607	.000	0.121
Opp.'s Delay	Opp.'s Delay	2	154	4.823	.009	0.059
Own Delay	QoE	2	154	19.310	.000	0.200

The perception of time in the VR condition (M = 3.22, SEM = 0.18) as rated on the S FSS-2 item was significantly more different to normal time compared to the condition without VR, where participants disagreed that time passed differently to normal (M = 2.90, SEM = 0.20). Results from the IPQ questionnaire revealed significant differences in each dimension. The item general presence showed a statistically significant difference between the condition with VR (M = 4.08, SEM = 0.16) and the condition without VR (M = 1.55, SEM = 0.26). Following this, the dimension of the spatial presence dimension was assessed significantly differently when using VR (M = 4.01, SEM = 0.12) than when using the condition without VR (M = 1.99, SEM = 0.20). The item measuring involvement is the next dimension that revealed significantly different values for the condition with VR (M = 3.36, SEM = 0.14) in comparison to the condition without VR (M = 1.42,

SEM = 0.13). Last but not least, the dimension of experienced realism was rated with a statistically significantly different value for the condition with VR (M = 2.87, SEM = 0.14) compared to the condition without VR (M = 1.94, SEM = 0.17). Lastly, when rating social presence with the NMQ questions, results indicated a significant difference for the item perceived comprehension when comparing the condition with VR (M = 2.72, SEM = 0.12) and the condition without VR (M = 2.42, SEM = 0.10).

When it comes to the judgment of the preferred system, overall, 24 people (66.6%) have selected the set-up with VR to be preferred, out of which 16 participants (44.4%) also wanted to have an opportunity to communicate with another player. In contrast, eight participants (22.2%) preferred to have just VR and no conversation.

4.1.1.2 Flow

In order to compare different visualizations of breathing biofeedback in VR exergame, Study 4 was conducted having as base condition the one where participants were not using VR. According to the data, which can be seen in Table 4.2, there were statistically significant differences in perceived flow due to the conditions. The side-corrected pairwise comparison indicated a statistically significant higher feeling of flow for the conditions VR All with all breathing visualizations shown at once (M = 3.966, SE = 0.112) compared to the condition No VR outside of VR (M = 3.367, SE = 0.135).

4.1.1.3 Skill as Correct Breathing

Furthermore, when comparing different set-ups for breathing biofeedback in VR exergame rowing, results from Study 4 have shown significant differences in performance determined by the percentage of correct breaths taken during a workout.

The observed breathing pattern showed certain results, as given in Table 4.2, such as how users could keep a correct breathing rhythm and suggested that some conditions influenced the percentage of correct breathing. In order to determine differences between conditions, the side-corrected pairwise comparison was performed. Participants in the condition No VR (M = 0.794, SE = 0.054) and VR Only (M = 0.831, SE = 0.044) maintained their breathing rate significantly better than participants in the conditions that were using VR with visualizations of biofeedback in the form of a chart as condition VR Chart (M = 0.722, SE = 0.064), as condition VR Lung with visualization of lungs in VR (M = 0.692, SE = 0.072), and in VR using all visualizations at once VR All (M = 0.701, SE = 0.070). Figure 4.1 shows the results and values for each scenario examined.

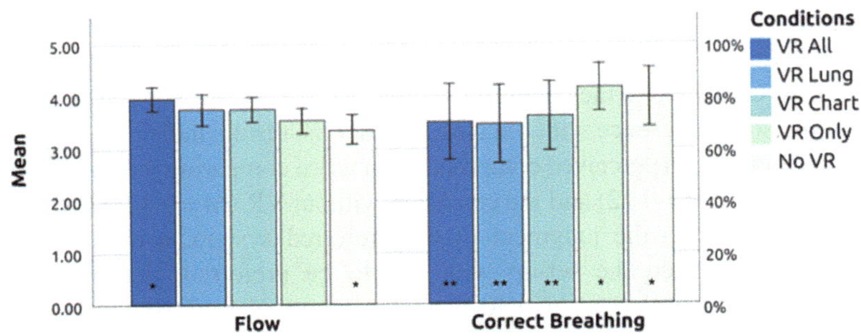

Fig. 4.1 Mean value of flow (FSS2) and correct breathing value over all conditions. Whiskers denote the standard error, while sign * indicates significant differences

4.1.2 *Impact of Network*

The impact of the network was explored for multiplier VR exergame with the scenario of the rowing race, with Study 3 focusing on the perception of users' delay. In such an example, the fairness of the game seen by users could be affected, influencing the full experience. Therefore, a feeling of presence and flow were measured by S FSS-2, IPQ, and perception of own and opponent's delay and overall feel for the quality of experience.

4.1.2.1 Presence and Flow

Participants' responses to the IPQ questionnaire demonstrated a statistically significant difference across all dimensions when the test participant's own network experienced a delay. The item's general presence was significantly lower when the player's delay was HIGH (M = 4.087, SEM = 0.190) than when it was MEDIUM (M = 4.978, SEM = 0.190) or LOW (M = 5.116, SEM = 0.155). Similarly, when their own network delay was HIGH (M = 4.222, SEM = 0.128) compared to MEDIUM (M = 4.687, SEM = 0.128) and LOW (M = 4.641, SEM = 0.105) for the player, participants evaluated the IPQ dimension for spatial presence lower. As an IPQ dimension, involvement showed a significant amount of variation depending on the delay setting that the participant had. The player engagement was significantly lower for HIGH delay (M = 3.630, SEM = 0.116) compared to a player on the game with LOW delay (M = 4.105, SEM = 0.095). Besides that, the experienced realism dimension of the IPQ showed a significantly lower result for HIGH delay (M = 3.152, SEM = 0.084) as compared to the condition in which the player's delay was LOW (M = 3.551, SEM = 0.068). The various delay choices had a considerable impact on flow, similar to how they did on presence. When the delay was adjusted to HIGH, the test subjects felt significantly less like they were in flow (M = 3.623, SEM = 0.086) than when it was set to MEDIUM (M = 3.966, SEM = 0.086) or LOW (M = 4.048, SEM = 0.070).

4.1.2.2 Judgment of Perceived Delay

During gameplay, participants had different delay sets in the simulation of the network on the player side and for the opponent. They were not informed where the delay occurred, but by viewing the game's visuals, they might see twitches or picture skipping depending on its intensity. If a delay occurred on the player's side, the complete environment was impacted, but on the opponent's side, just the opponent's representation was affected. Participants perceived their own delay to be significantly longer (M = 7.196, SEM = 0.322) during the HIGH delay condition than during the MEDIUM (M = 4.022, SEM = 0.322) or LOW ((M = 3.087, SEM = 0.263) delay conditions. However, no significant differences were found between the MEDIUM and LOW delay conditions. Furthermore, when their own delay was HIGH (M = 7.826, SEM = 0.338), participants rated their opponent's delay significantly higher than when their own delay was MEDIUM (M = 5.804, SEM = 0.338) or LOW (M = 5.565, SEM = 0.276).

In addition, the only significant difference that was observed to be dependent on the opponent's delay was in judging how long the opponent was thought to be delaying. Under the circumstances in which the opponent's delay was set to HIGH, the perceived opponent's delay was assessed significantly higher (M = 7.478, SEM = 0.338) than it was under the situations in which the opponent's delay was set to MEDIUM (M = 5.630, SEM = 0.338) or LOW (M = 5.913, SEM = 0.276).

4.1.2.3 Overall QoE

The amount of delay set-up for the player during gaming impacted the QoE's rating, which revealed significantly diverse outcomes depending on the level of delay. The overall average rating of the QoE is shown in Fig. 4.2. It can be shown that the QoE was evaluated significantly lower when the own player's delay was HIGH (M = 4.609, SEM = 0.318), compared to MEDIUM delay (M = 6.804, SEM = 0.318), or LOW delay (M = 7.174, SEM = 0.260), where it was reported to be the biggest. However, no significant difference was observed for the opponent's delay, and QoE was judged acceptable for all situations when the player's own delay was minimal.

4.2 Discussion

To address the question of how VR influences motivation or engagement (RQ1.1), it is necessary to look at the factors shown by the results and see how they relate to SDT, the sense of presence, and flow theory.

Given that SDT as a theory of motivation takes into account psychological needs like competence, autonomy, and relatedness, it is also known in gaming that the user's desire to do something again is tied to a sense of presence [133]. That means

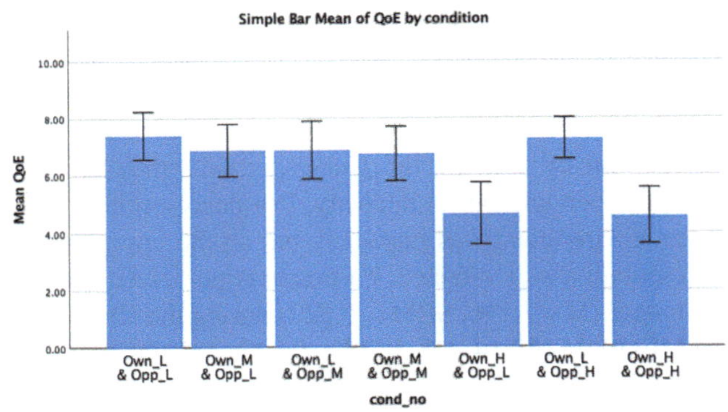

Fig. 4.2 Average rating of the user's QoE for the different conditions of the player's (Own) and opponent's (Opp.) delay (L: 30 ms, M: 100 ms, and H: 500 ms). Whiskers denote the 95 % confidence interval

that a sense of presence is linked to intrinsic motivation. Also, both autonomy and competence make players more interested and motivated. That is because competence as a need for challenge and a sense of achievement [238] affects the balance between boredom and fear during the game, which is related to the theory of flow [112, 137]. When users are in a state of flow, they are motivated to do something simply because it interests them. Such effects have been shown to be true, especially for athletes whose intrinsic motivation and flow are positively linked.

The results showed that the addition of a virtual environment increases feelings of general presence, spatial presence, involvement, and experienced realism and that all dimensions of the feelings of presence IPQ measurement were rated significantly higher when compared to the control set-up. That suggests that activity in a virtual environment may be more interesting and realistic for users, and it is aligned with results from gamification research [193]. Furthermore, users reported that the VR system provided them with a significantly higher level of perceived comprehension, which could be attributed to a more detailed visual representation of the environment and the opponent himself. All of these findings were confirmed by the fact that the majority of participants chose the VR set-up as the preferred system, and it could be assumed that such an experience would be repeated with VR. Given that increased presence affects motivation and UX, an effect that is also occurring in online games [239], it is reasonable to conclude that the VR device positively influences UX.

Regarding the reported flow based on S FSS-2, the analyses have shown differences in flow in two studies (Study 1 and Study 4). In a multiplayer setting, participants perceived time significantly different compared to the real-time passing in the system when no VR was used. Forgetting about the time when gaming or reporting about the difference in passing the time is an effect when users are in the flow of activity (as previously explained in Sec. 2.2.2), and in this case, it might have happened because VR offered users more than just an activity. A similar effect has been reported in single player, where the task of the game was not the race but the exercise in breathing for it. When having VR setting with visualization of breathing for biofeedback, users have reported a significantly higher flow rate compared to the setting with no VR. Knowing that flow happens when users are not overwhelmed with tasks nor task is too easy for them can confirm that the task level was correctly set for participants.

Users of the ideal VR rowing exergame should be able to focus on the game and develop the right breathing rhythm without interfering with the flow of the VR experience. However, the right breathing rhythm has been reduced for VR representations, perhaps due to the participant's visual distraction. Even though the condition without the VR had the best breathing rate, participants felt significantly more in the flow state with visualizations in VR when they were shown how to breathe. These results do not align with the flow theory [136], but it is possible that the users were distracted by the graphics from the breathing synchronization task due to not good enough design of content, which is a hypothesis that could be investigated further.

Furthermore, to determine whether delay influenced the experience when playing VR series games (RQ1.2), the sense of presence and flow as UX parameters were examined, and additionally, participants were explicitly questioned if they sensed delay. Because the scenario for this study is strongly dependent on delay since it is an exergaming race, participants were also asked to evaluate if the delay happens on their or the network of opponents. The results of experimental research revealed that network delays could significantly impact the feeling of presence, flow, and overall experience ratings, confirming similarities with ITU online gaming work [186]. Furthermore, participants could not always distinguish whether the delay was on their or an opponent's side. As delay increases, both the sense of presence and flow decrease. Given that the result of network delays was manifested in quality degradation, such as image jitters and image skipping or even an unfair final outcome of the game, such results could have been expected and match with online video game results [240]. Such a fragmented picture presentation, especially when that image surrounds the user, appears unrealistic because the world is never frozen in front of a person in real life. The fact that such a presentation is unrealistic is also supported by the results, which show that the delayed system was rated significantly less realistic than the system with minimal delay. Furthermore, image and information delays can cause users to lose trust in data reliability, making such a system unreliable.

It is also interesting to note that users could not always tell if the problem with the network was on their side or the opponent's side. The results demonstrated that a player with a network delay cannot appropriately evaluate the situation and believes that the opponent player is late. That is, when a player has a bad network, he assumes that the quality is bad not only because of own settings but also because the opponent has network problems. Finally, the overall experience is likely to be affected by network delay, and player network delay significantly impacted QoE. However, the difference in QoE between LOW and MEDIUM delay is not significant, and while a MEDIUM delay of 50 ms is longer than recommended [160], this experiment did not result in a lower quality score. Nevertheless, if we consider the content of the game, that in races the player did not have real interaction or communication

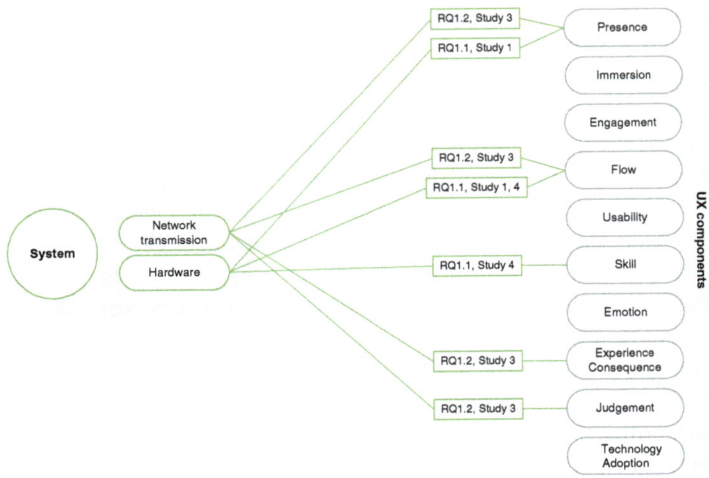

Fig. 4.3 Illustration of connections between System IF, particularly Network transmission and Hardware, to UX components where significant effects from a particular study contribute to answering RQs

with the opponent and that important information was the opponent's position, such results could be explained.

Overall, results have demonstrated significant effects on several components of UX, as illustrated in Fig. 4.3. Presence and a sense of flow, in particular, have been significantly addressed by both subfactors, Hardware and Network transmission, but in opposite ways. As a result, as long as network delays are minimal, it is reasonable to predict that user motivation will increase with the use of VR systems. Otherwise, motivation and overall sense for the QoE will be significantly reduced, with users not correctly determining if problems are occurring in their or other players' systems.

4.2.1 System Design Guidelines

Based on the results and observations from the studies, the following suggestions can be concluded regarding System IFs when designing and developing VR serious games:

- Consider using VR solutions for serious purposes, as they increase the feeling of presence and users feel in flow when solving some task. Motivated users want to solve the tasks.
- Be aware that VR solutions can sometimes also decrease performance with a task. Ensure that environment is not overwhelming and/or graphically overloaded.

- Mind that network errors, such as delay, will decrease users' feeling of presence, flow, and, therefore, overall motivation and experience. Monitor and inform users when problems in the network are happening.

4.3 Summary

This chapter has presented results from the research focused on the effects that System IFs may have on the UX and how those effects are manifested. Specifically, the purpose of this chapter was to investigate whether or not the VR set-up has an effect on the motivation for serious VR gaming (RQ1.1) and, in the case of the multiplayer scenario, what impact the network latency has on the user's impression of the experience (RQ1.2). Therefore the particular focus was placed on Hardware System IF and network-related System IF that were part of the research carried out in Studies 1, 3, and 4.

Results, divided by the influence of each subfactor device and network delay, report significant effects found with data analyses. In particular, using a VR device was shown to influence the feeling of presence, flow, and skill (measured by the performance of correct breathing for biofeedback), while network delay influenced the feeling of presence, flow, and overall QoE and judgment of where delay and problems were occurring.

Furthermore, results show that both factors are significantly influencing the feeling of presence and flow. Therefore, effects are discussed in relation to SDT and the theory of flow and presence. Since the user's desire to do something again is tied to a sense of presence, which means that a sense of presence is related to motivation that comes from within. Also, users are motivated to do something that interests them when they are in a state of flow. Consequently, it can be observed from the findings that VR devices have the potential to raise the user's motivation, whereas delays in the network may have the opposite effect.

Finally, Fig. 4.3 summarizes all the findings, indicating that System IFs have significantly influenced several UX components.

Chapter 5
Content IF

5.1 Results

This chapter aims to investigate how serious games' content for VR technology influences UX. Content is a subfactor of the System IFs, and in the current overview for VR services, it is not included as a separate category but as a subcategory. Although Content is still considered a component of System IF, in gaming, it is further subdivided into Narrative, User Representation, Environment Design, and Supported Interactions. Such elements may also influence the experience of serious games in VR, which is what this chapter wants to investigate.

As VR applications try to immerse the user in the world and show the virtual environment as realistically as possible, the first-person view is commonly used. So such setting was also chosen for this research, and the User Representation factor will not be examined. Furthermore, since the experiments were done with the aim of researching specific factors with some very specific requirements and needed repetitions, the narrative of the games is another factor that is out of scope for this work. However, both factors were controlled and did not change during either experiment.

The study focuses on the Interactions and Environment UI Design aspects for serious games in VR. The goal was to specifically study how the design of virtual environments and interactions influences the UX, particularly emotions and the feeling of presence for users (RQ2.1), as well as how different positioning and complexities of UI elements in VR affect usability (RQ2.2). As a result, three experiments were conducted as Studies 2, 5, and 6 (a summary of the methodology is included in Table C.2), and the results are given later in this chapter.

A repeated test ANOVA was used to find statistically significant differences in order to evaluate the effects of different conditions. Further on, the Wilcoxon signed-rank test was done to find out significant differences in preference of interaction types.

T. Kojić, *User Experience for Serious Games in Virtual Reality*, T-Labs Series in Telecommunication Services, https://doi.org/10.1007/978-3-031-75530-9_5

Table 5.1 Influence of different interactions represented as hand tracking or controllers with different visualizations of hands and controllers for two different tasks using grabbing or typing interaction, on arousal (SAM_A), valence (SAM_V), and dominance (SAM_D) measured with SAM, and feeling of presence (PRES) and realism (REAL) measured with IPQ

Parameter	Effect	df_n	df_d	F	p	η^2_G
Task	SAM_A	1	20	10.35	.005	0.35
Task	PRES	1	20	5.30	.033	0.22
Task	REAL	1	20	4.90	.039	0.21
Interaction	SAM_A	3	20	5.25	.003	0.22
Interaction	SAM_V	3	20	20.08	< .001	0.51
Interaction	SAM_D	3	20	11.43	< .001	0.38

Table 5.2 Wilcoxon signed-rank tests were used for post hoc examination of significantly different effects of ranking for all interaction types: controllers that show both hands and controllers (C_H), controllers that show only hands (C_OnlyH), controllers that show just controllers (C_noH), and hand tracking (HTrack)

Interaction	df_n	df_d	Z	p
C_H and C_OnlyH	3	20	−3.079	.002
C_H and C_noH	3	20	−3.162	.002
C_H and HTrack	3	20	−2.844	.004

Tables 5.1, 5.3, and 5.4 provide an overview of the statistically significant ANOVA tests, while Table 5.2 reports results of the Wilcoxon signed-rank tests. Additionally, Table C.1 shows for all conditions mean values that participants have selected as the best settings of text parameters of size, distance, and color contrast. In the following sections, all of the significant influences reported will be discussed and explained.

5.1.1 Influence of Interactions

5.1.1.1 Emotions

In order to examine the influence of different interactions, in Study 6, the two tasks that were used were grabbing and typing, as they require different hand interactions. Interaction types compared were those using controllers with different hand visualizations and hand tracking. Results report that there were statistically significant differences in perceived arousal, presence, and realism that were assessed by SAM and feeling of presence measured by IPQ.

Results have discovered that the varied situations in which participants performed various task types, such as grabbing or typing, affected how the users perceived the SAM dimensions, presence, and feeling of realism. When it comes to the reported emotions, participants perceived typing task as significantly more arousing (M = 3.83, SE = 0.21) compared to the grabbing task (M = 3.54, SE = 0.22). How-

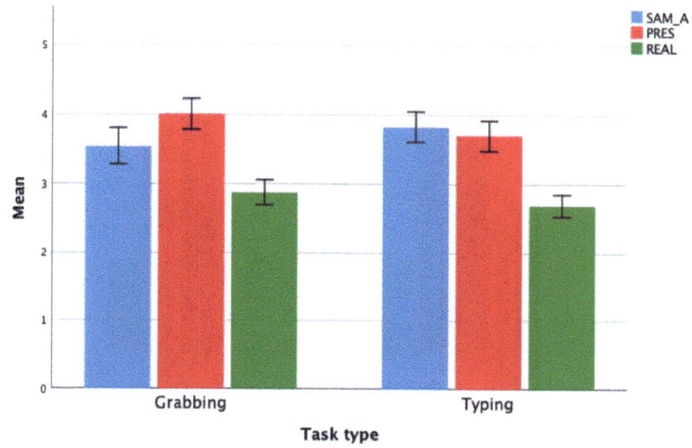

Fig. 5.1 Mean values of emotional response measured by SAM for arousal (SAM_A), presence (PRES), and realism (REAL) over all participants for all task types. Whiskers denote 95% confidence intervals

ever, for the feeling of presence and realism, the participants reported experiencing a significantly increased feeling of presence when performing the task in which they were required to grab the balls (M = 4.00, SE = 0.15) in comparison to the task in which they were had to type (M = 3.70, SE = 0.17). In a similar manner, the grabbing task produced a significantly stronger sense of realism (M = 2.88, SE = 0.14) in comparison to the typing task (M = 2.69, SE = 0.14). All effects regarding different interaction task types are shown in Fig. 5.1.

When the effects induced by the various interaction methods with different visualizations of controllers or hand tracking were compared, statistically significant differences in emotional responses were identified for all three aspects of the SAM questionnaire, as demonstrated by Fig. 5.2. Hand tracking was rated significantly lower for arousal (M = 3.38, SE = 0.21) compared to the arousal of interaction with the controller where only hands were visualized (M = 3.83, SE = 0.21) and higher for valence (M = 2.40, SE = 0.20) compared to all other interaction types with controllers visualized either as only controllers (M = 1.65, SE = 0.14), only hands (M = 1.48, SE = 0.12), or with visualization of both (M = 1.58, SE = 0.16). However, it is interesting to note that the dimension of dominance was reported to be significantly lower for hand tracking (M = 3.13, SE = 0.16) compared to visualization of both controllers and hands (M = 3.85, SE = 0.14), visualization of only controllers (M = 3.75, SE = 0.16), or visualization of only hands (M = 3.70, SE = 0.14).

The participants were asked to rank their favorite manner of interaction types, going from best (4) to worst (1), and the findings showed that there were statistically significant differences across the options. A side-corrected pairwise comparison showed that interaction types with controllers where both hands and controllers were visualized were assigned a significantly higher rank (M = 3.30, SE = 0.47) compared to all other conditions. These other conditions included visualizations of

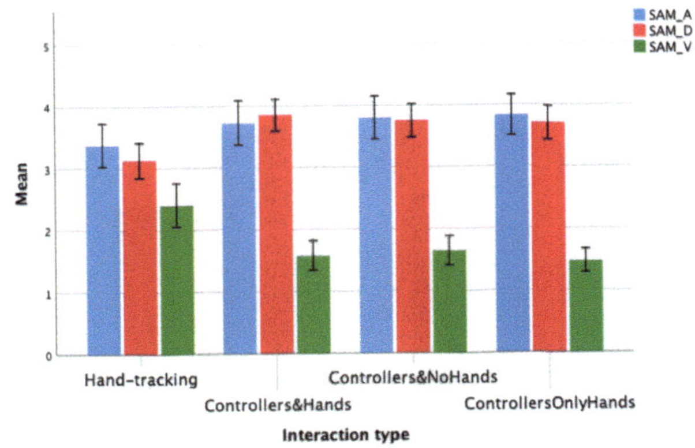

Fig. 5.2 Mean values of the SAM dimensions: arousal (SAM_A), valence (SAM_V), and dominance (SAM_D) over all participants for all interaction types with controllers (visualized as both controllers and hands, only controller, only hands) and hand tracking. Whiskers denote 95% confidence intervals

only controllers (M = 2.35, SE = 0.93), controllers with visualizations of only hands (M = 2.25, SE = 1.16), and hand tracking (M = 2.10, SE = 1.37).

5.1.2 *Influence of UI Design*

5.1.2.1 Judgment by Angular Size

To explore how different text settings affect readability in VR, Study 5 was conducted by focusing on the text parameters of font size and text distances, which will be referred to as angular size. The distance independent millimeter (dmm), which is Google's unit for angular size, is used to define angular size, where 1 dmm is equivalent to 1 millimeter in height when seen from a distance of 1 meter. Angular size values were determined by participants for the optimum text parameter settings differed by adjusting the length of the text: 2 words (short), 21 words (medium), 51 words (long), and device (Oculus Go and Quest). The different settings of text parameters have reported significant differences in dmm and emotional responses of participants, shown in Fig. 5.3.

When comparing values for dmm that users reported as preferred and those that users selected for negative readability, results show that for each condition, dmm values for the negative setting are significantly higher than for the positive task, as shown in Fig. 5.4. This is the case both for Oculus Go with short (M = 297.12 SD = 691.48), medium (M = 203.51 SD = 389.75), and long (M = 203.51 SD = 389.75) text and for Oculus Quest with short (M = 404.60 SD = 866.13), medium (M = 253.14 SD = 365.15), and long (M = 169.52 SD = 235.95) text (Table 5.3).

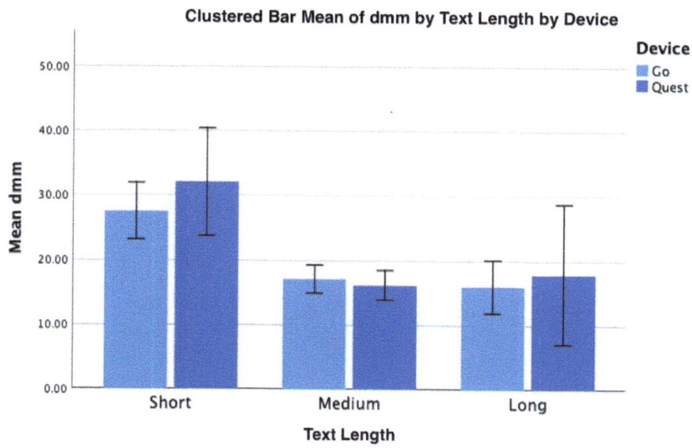

Fig. 5.3 The average value of the angular size, measured in dmm, when selecting the optimal text parameters for reading in virtual reality, taking into account the length of the text display and the kind of VR device used. Whiskers denote 95% confidence intervals

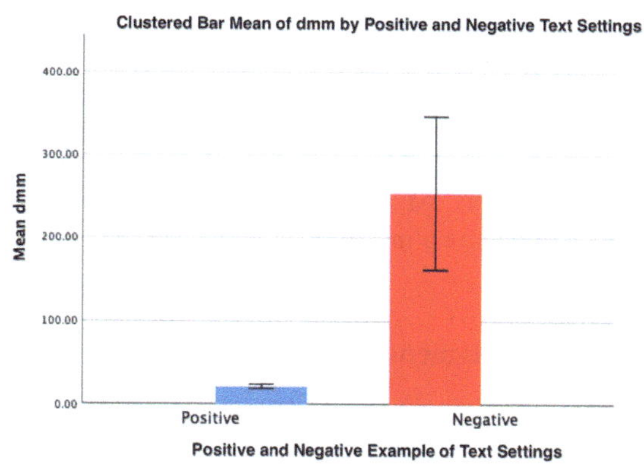

Fig. 5.4 The average values of the angular size, measured in dmm, for each of the two tasks, which consisted of selecting the most optimal (positive) and the least (negative) text properties for reading in VR. Whiskers denote 95% confidence intervals

Table 5.3 Influence of text length (Length) on angular size (dmm) as a combination of font size and distance and SAM dimension of dominance (SAM_D), in different tasks, different HMD devices (Device), and task if users had to select the best settings (Pos) or the worst (Neg)

Parameter	CoVar	Effect	df_n	df_d	F	p	η_G^2
Length & Pos		dmm	1	22	9.89	.005	0.32
Length & Neg	Device	SAM_D	1	22	6.14	.022	0.22

5.1.2.2 Emotions

Regarding the emotions participants reported while reading text in the worst possible settings, there have been significant differences depending on the text length. For reading long text, participants feel significantly more in control while using the device Oculus Quest (M = 3.13, SD = 1.52) compared to the same long text while using the device Oculus Go (M = 2.50, SD = 1.40). Instead, for medium

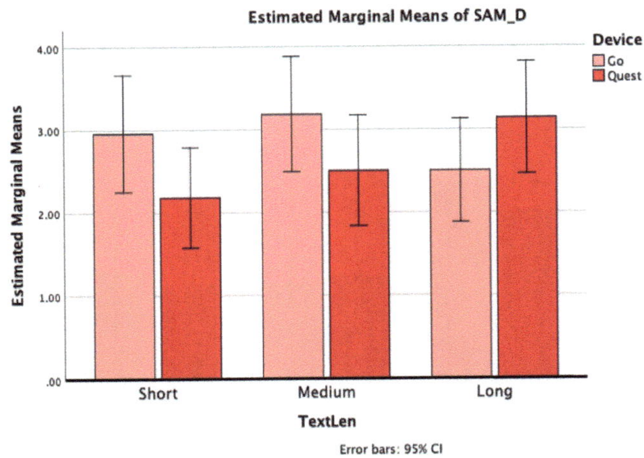

Fig. 5.5 The average value of the SAM dominance dimension for the job of selecting the worst text characteristics to consider while reading in VR, in terms of the type of VR device that was used. Whiskers denote 95% confidence intervals

and short text, the effect was the opposite. The reported impression of dominance was significantly greater with the device Oculus Go for short (M = 2.95, SD = 1.58) and medium text (M = 3.18, SD = 1.56) compared to when using Oculus Quest with the same short (M = 2.18, SD = 1.36) and medium (M = 2.50, SD = 1.50) text. The effects are shown in Fig. 5.5.

5.1.2.3 Judgment by Contrast Ratio

In order to further explore text parameters in Study 5, the contrast ratio was calculated by comparing the color of the text to the color of the background in a virtual environment. For the preferred settings of contrast radio, all mean values are given in Table C.1 in the Appendix. Even while the length of the text did not significantly affect the contrast ratio or emotional response of the participants, it is noteworthy to report the frequency of reporting values, as it is shown in Fig. 5.6 to notice the distribution of contrast, in particular for the highest contrast (21:1), where it was rated as the second worst option for reading by participants. These results led to significantly different mean values for a contrast ratio for all text lengths and devices (see Fig. 5.7).

5.1.2.4 Usability as Readability and Support

Furthermore, to explore the influence of UI design, in Study 2, there have been four different visualizations of metrics for the VR rowing exergame, where participants had a task to explore and report on their experience. Visualizations have deferred in

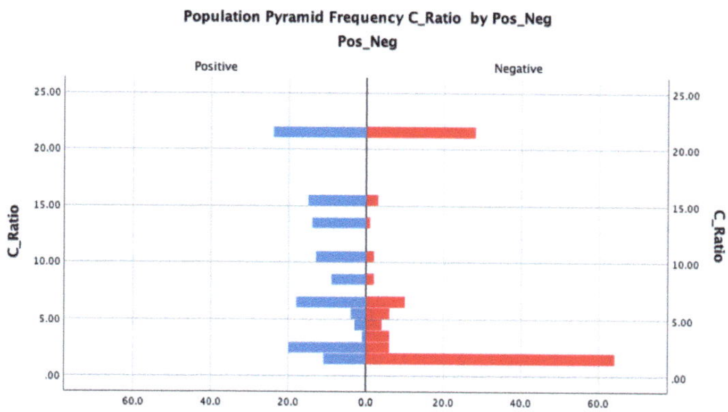

Fig. 5.6 Frequency of values of contrast ratio between the color of text and background for both tasks that participants chose as the optimal (positive) and the worst (negative) reading experience in VR

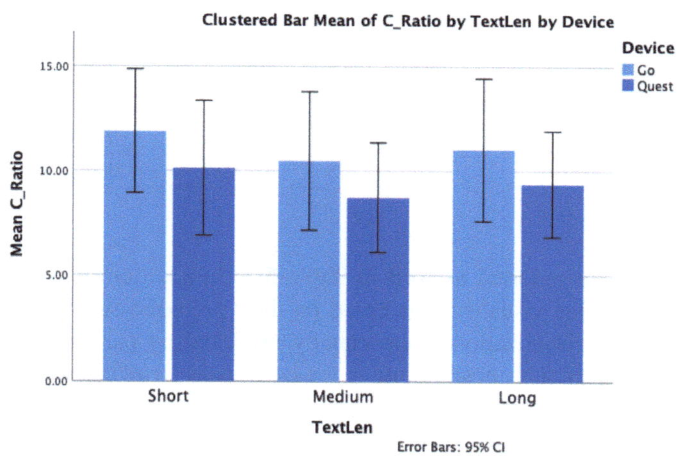

Fig. 5.7 The mean result of the contrast ratio between the color of the text and the background for the task of selecting the optimal text parameters for reading in VR, taking into account the amount of text that was shown on the VR device that was being used. Whiskers denote 95% confidence intervals

the position where information was displayed to the user and the complexity where more or less information about the workout was displayed.

The complexity level affected how easy it was for a user to read metrics, with digital visualization (M = 5.67, SD = 1.34) being significantly easier to read than gamified visualization (M = 5.02, SD = 1.41). Figure 5.8a displays these results. Furthermore, the positioning of the metrics influenced the perception of received support within the gameplay. As shown in Fig. 5.8b, the level of support was significantly higher when metrics were displayed on a boat moving next to them (M = 6.15, SD = 0.8) compared to metrics displayed in a cockpit (M = 5.87, SD = 0.95). Both

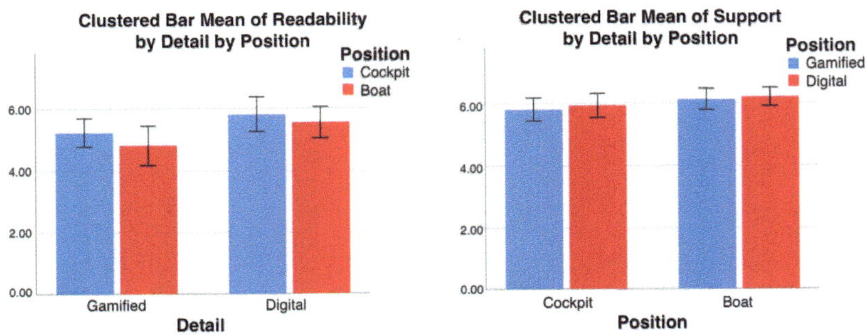

(a) Readability (b) Support

Fig. 5.8 (**a**) Mean value for readability of all participants for the conditions with gamified and digital visualization split by the position of metrics. (**b**) Mean value for perceived support of all participants for the conditions with the position of visualization as the cockpit and in a boat split by complexity. Whiskers denote the standard error

Table 5.4 Effects of different complexity and positioning of elements in UI for VR rowing exergame on readability (read) and support to the user during a workout (support)

Parameter	Effect	df_n	df_d	F	p	η^2_G
Complexity	read	1	25	6.71	.016	0.21
Position	support	1	25	4.88	.037	0.16

of these effects are reported as well in Table 5.4. In addition, participants were asked to rank various visualizations based on their preferred settings. The results revealed that two visualizations were equally preferred by participants: gamified visualization metrics with positioning as the cockpit in a rowing scull (33%) and digital visualization of metrics presented on a boat (33%). Other visualizations that were less preferred were gamified visualization on a boat (18.5%) and the cockpit with digital metrics visualization (14.6%).

5.2 Discussion

To answer the question of how the content of a virtual environment affects users in serious VR games (RQ2.1), the factors of the visual design of the environment itself and the interaction with a given environment were particularly considered.

Firstly, the influence of the virtual environment was evaluated by the influence of UI design and, in particular, text in VR. Since the text is often, in reality, surrounding people, as it is used to mark and explain things when creating an environment, it is very important not only to position objects correctly (as explained in Sect. 2.1.2.4) but also to correctly display the text that goes with them in a virtual environment.

Therefore, this work focuses on experimentally exploring text reading parameters for users in VR, with different text lengths, using two different devices. As a task, participants were asked to choose the best and worst possible settings to create the most comfortable reading experience.

Regarding the UX of reading in VR, the results show that the text's length and the device on which it is displayed significantly affect how users perceive the experience. The results show that the length of the text affects emotions and the sense of dominance measured by the SAM dimension. Users reported having greater control while reading text on the Oculus Quest device compared to reading the same text on the Oculus Go device if the text is set according to the selected worst parameters. This may be due to technical differences in the devices, such as resolution [180]. Still, interestingly, no difference was found when selecting the best parameters, but the devices affected the reading with negatively set text parameters.

Concerning text, users do not prefer a larger font just because the virtual environment provides a bigger screen size, that is, even though VR allows more space to be used. However, as expected based on Google dmm guidelines [165], there is a significant difference in size between preferred and unacceptable text size values. It can be noticed that the average values for the negative readability examples were much higher because the users opted to create a very angular size option for the negative option, i.e., they chose a large font for the text displayed at a very short distance. If these values are compared to standards such as WCAG2.1 [169], even negative examples will be correct for the standard because only a minimum text size is proposed. However, the results presented in this research represent the idea that it should be important to also define and standardize the maximum values for angular size.

The results show a similar effect with contrast ratio; although the WCAG2.1 standard [169] suggests guidelines only for conditions that must be minimally met, the study shows interesting results on the ratio between text contrast and background. The contrast of the maximum value, defined as 21:1 and representing black text on a white background or white text on a black background, was not preferred and was rated by 1/3 of users as the worst possible setting for reading text in VR.

Correspondingly, in the case where participants had to choose the best possible text settings, the mean contrast ratio results did not differ significantly for different text lengths, nor did they affect UX differently, and were at least 7:1 in all conditions, as required by WCAG2.1 for AAA level. However, these results are related only to the grayscale contrast ratio, and although the colors can be reduced to a greyscale ratio, further standardization would require further investigation. In any case, creating design guidelines with only the minimum required contrast ratio is not enough because when designing an environment with a very bright background, since the VR device screen is very close to the eye, it can produce too much light, which users associate with a negative experience.

Furthermore, to investigate the impact of content depending on the interaction tasks performed by the user (RQ2.1), the results showed that different types of tasks would affect how users experienced the VR serious game. The two tasks assigned to users were based on very common actions that one can relate to office work and tasks that people perform: typing and moving objects. While moving an object, the user had to grab the object and move it to another place, and results have reported that the users with that interaction experienced greater presence, higher realism, and lower excitement compared to typing. This result can be attributed to the fact that the typing task uses a virtual keyboard, although the keyboard itself could be connected as one of the inputs to VR, and it seems that capturing virtual objects is considered more natural and realistic by users. However, this can be the case only until there are no mismatches between virtual and real objects [241]. Even though, in this experiment, no real objects were used besides controllers, even just the slight mismatch between users' real hands and virtual position of them for hand tracking could have had an influence.

The type and visual presentation of the interaction affected all SAM dimensions (valence, arousal, and dominance), meaning that it affected the emotions that the user experiences during the experience. The results show that participants felt less excited, had a more positive experience due to higher valence, and felt less dominant during hand tracking compared to all types of interactions involving controllers. A higher valence for hand tracking may relate to the fact that users like the experience, while a lower experience of dominance could be explained because interaction with hand tracking leads to a lower sense of control or precision [79], meaning that these results are in line with assumptions from research in interactivity, which in the case of hand tracking has been focused the most on the precision of tracker, sensors, and devices (see Sect. 2.1.2.5).

Finally, participants ranked the type of interaction with controllers visualized as both controllers and hands as preferred over all other types of interaction. The latter effect could be a combination of the usage and precision of controllers and that most tasks are currently quite developed for selecting with using controllers. It seems that users still prefer to visualize what is realistic when it comes to visualization and that users use not only the controller but also their hands when interacting.

When it comes to research on UI design and how elements can be positioned within a virtual environment concerning the position and complexity of these elements (RQ 2.2), various metrics within VR rowing exergaming were visualized. The goal of the study was to determine how the positioning of metrics (as a cockpit at the front of the rowing scull or in the following coach boat) and the complexity of the rowing metrics (as digital numbers or as gamified visualization) influence UX.

With the results presented, it has been shown that the positioning and complexity of UI, such as metrics in rowing, affect the experience in VR exergaming. In particular, users rated the sense of support and readability differently for different visualizations. Thus, digital visualization was rated as significantly more readable than gamified metrics, which can be explained by the fact that the digital version displayed only numbers while the gamified version displayed a speedometer. When it comes to the different perceived levels of support for different visualizations, the reason behind this might be because the task was to focus on maintaining speed; users were able to read numbers more easily than following a speedometer, an effect that could be expected in reality.

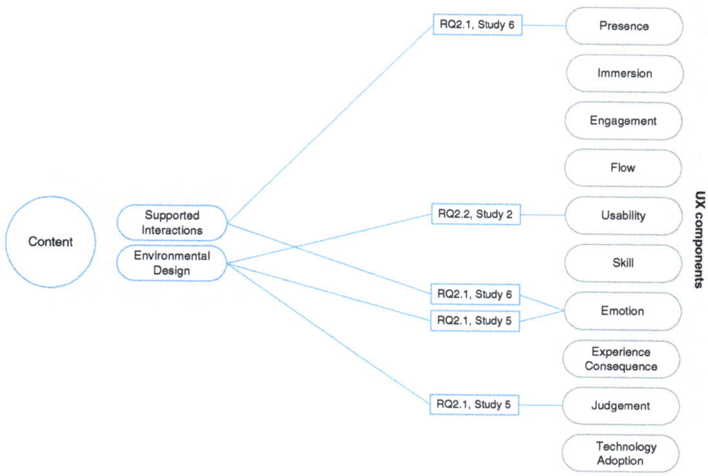

Fig. 5.9 Illustration of connections between Content IF, particularly Supported Interactions and Environmental Design, to UX components where significant effects from a particular study contribute to answering RQs

Overall, two opposing visualizations were selected as preferred by the participants, namely the cockpit with gamified visualization and the digital visualization of the metrics in the boat. The results show a trend that users prefer different visualization depending on the position of metrics. However, in order to be able to make a final decision as to which design is better for a user, it is necessary to consider other aspects with different covariants, which will be explained further under Human IF. Furthermore, since both metric visualizations have some advantages and disadvantages compared to each other, the final design decision must be customized to the user, as proposed with user-centric design [13] or flexible to the possibilities of different settings based on the profile of a particular user.

Finally, as shown in Fig. 5.9, the findings have shown some significant influences that have been measured for aspects of UX. For both of the researched Content subfactors, Supported Interactions and Environmental Design, significant influences on Emotions have been reported as SAM dimensions. Depending on where and how the elements are presented in UI, as well as different interaction tasks, users feel more or less excited or in control of the experience. Furthermore, a feeling of presence was analyzed separately, and it seems that presence and realism are significantly influenced by subfactor Supported Interactions depending on the interaction tasks. When it comes to the different Environment Designs, the overall usability and experience are influenced, in particular readability or support that users feel with particular designs. Therefore, this work reports on minimal and maximal values for angular size and contrast ratio of text as starting points for standardization, as well as design guidelines that can be used when developing content for VR serious games.

5.2.1 *Content Design Guidelines*

The following guidelines are proposed based on the results and conclusions of user studies, with an emphasis on the design and development of the content of VR serious games:

- Mind that content size and interaction type can influence users' emotions and how they feel. Until there are new standards with both minimum requirements and limitations of maximal values, be aware that creating too large elements with too big contrast can decrease UX.
- Creating content for some tasks can have more than one visual solution, so try to enable flexibility for users to choose between different views.

5.3 Summary

This chapter highlighted study findings on the influence that content creation may have on UX and how various visualizations are perceived by users. The goal of this chapter, in particular, was to study whether or not virtual environments have an influence on users' feelings of presence and emotions (RQ2.1), moreover dependent on the positioning and complexity of UI elements in VR serious games (RQ2.2). As a result, special emphasis was put on the subfactors Environmental Design and Supported Interaction, explored during the studies conducted as Studies 2, 5, and 6.

After analyzing the data from user studies, the results report that how UI is designed and how users interact within a virtual environment significantly affect UX. As of SAM dimensions for both subfactors, Supported Interactions and Environmental Design, significant differences in results on Emotions have been observed. Depending on where and how the elements are shown in the UI and the various interactions they perform, users may feel more or less excited or in control of the experience. Furthermore, the experience of presence was explored, and the findings reveal that, depending on the interaction tasks, the subfactor Supported Interactions significantly influences the feeling of presence and realism.

Additionally, when analyzing results for readability of UI, interesting results were found about the visualization of text in VR. It is worth noting that previous guidelines have specified values for contrast ratio and font size as the absolute minimum that must be fulfilled. However, when users in this research were asked to pick the worst possible conditions for reading in virtual reality, participants rated angular size significantly higher compared to the condition when they were asked to choose the best angular size for reading in VR. Regardless of having a display and virtual environment that could handle big fonts, it was not something that users preferred regarding readability. Likewise, users did not prefer contrast ratios with maximal values, and some rated it as the worst set of text settings.

Lastly, a summary of all the results can be seen in Fig. 5.9, which suggests that Content IFs have influenced several different UX components.

Chapter 6
Context IF

6.1 Results

This chapter intends to study context settings in which VR serious games may be used, as well as the effects on users that such a VR setup can have. Because this research is based on lab-based research, most subfactors are treated as controlled variables because of the limits of in-lab research and ethical approaches to user studies. As a result, the importance of social factors was emphasized in this work.

The goal is to investigate how social environments impact the social acceptability of VR as a tool and users' emotional responses (RQ3.1), including how the opportunity to have conversations influences users' experiences (RQ3.2). User studies that have these questions in focus were Study 1 and Study 8 (a summary of details about studies is shown in Table C.2), and all significant results are presented in the rest of this chapter. An overview of all significant effects that will be explained in the following sections is given in Tables 6.1 and 6.2 from the performed analyses that was ANOVA.

6.1.1 Influence of Social Environments

In order to explore the influence of different social environments depending on the number of people and distance from the user when users are playing a static or a dynamic game in Study 8, different scenes were presented to the user in a virtual environment, and different components of UX were measured with questionnaires UEQ-S, SAQ, IPQ, and SAM.

T. Kojić, *User Experience for Serious Games in Virtual Reality*, T-Labs Series in Telecommunication Services, https://doi.org/10.1007/978-3-031-75530-9_6

Table 6.1 Results show significant influences of Social Environments (Environment) and degree of interactivity (Interactivity) on UEQ-S Overall and Hedonic quality scores (UEQ-S_Overall, UEQ-S_Hedonic), Social Acceptability Questionnaire scores (SAQ_Public_VR, SAQ_Interaction, SAQ_Isolation, SAQ_Privacy, SAQ_Safety), IPQ general dimension (IPQ_G1), and SAM dimensions of valence and arousal (SAM_V, SAM_A)

Parameter	Effect	df_n	df_d	F	p	η^2_G
Env.	UEQ-S_Overall	3	81	2.785	0.046	0.093
Env.	SAQ_Public_VR	1	27	14.317	0.001	0.347
Env.	SAQ_Interaction	1	27	8.647	0.007	0.243
Env.	SAQ_Isolation	3	81	2.829	0.044	0.095
Env.	SAQ_Privacy	1	27	6.533	0.017	0.195
Env.	SAQ_Safety	1	27	11.913	0.002	0.306
Int.	IPQ_G1	1	27	10.075	0.004	0.272
Int.	UEQ-S_Overall	1	27	6.754	0.015	0.200
Int.	UEQ-S_Hedonic	1	27	6.869	0.014	0.203
Int.	SAQ_Safety	1	27	7.717	0.010	0.222
Int.	SAM_V	1	27	15.707	<0.001	0.368
Int.	SAM_A	1	27	13.106	0.001	0.327
Env. and Int.	SAM_A	3	81	2.948	0.038	0.098

Table 6.2 Effects of *VR* and *DTC* on IPQ *general presence G1*, and NMQ dimensions *perceived emotional contagion PE-EC*, *perceived comprehension PE-C*, and *behavioral interdependence BI*

Parameter	Effect	df_n	df_d	F	p	η^2_G
DTC	G1	1	35	8.49	0.006	0.19
DTC	PE-EC	1	35	13.97	0.001	0.28
DTC	PE-C	1	35	48.61	<0.001	0.58
DTC	BI	1	35	10.22	0.003	0.22

6.1.1.1 Experience Consequence by UX

Results have reported significant differences regarding how social environments influenced overall perceived UX measured by UEQ-S also depending on the level of interactivity in environments.

The findings revealed that participants reported a considerably lower UEQ-S Overall score when the static interaction was used (M = 1.175, SE = 0.148) compared to when they used the dynamic one (M = 1.417, SE = 0.152). For instance, the UEQ-S Hedonic component demonstrated a significant difference between the static interaction (M = 0.603, SE = 0.237) and the dynamic interaction (M = 1.016, SE = 0.223).

There is a relationship between the independent variable Social Environments and the dependent variables User Experience and Social Acceptability. Participants have reported a significantly better UEQ-S Overall score for the Social Environment with one distant person (M = 1.424, SE = 0.135) compared to the Social Environment with a few close persons (M = 1.205, SE = 0.161), as shown by the results. These results are shown in Fig. 6.1.

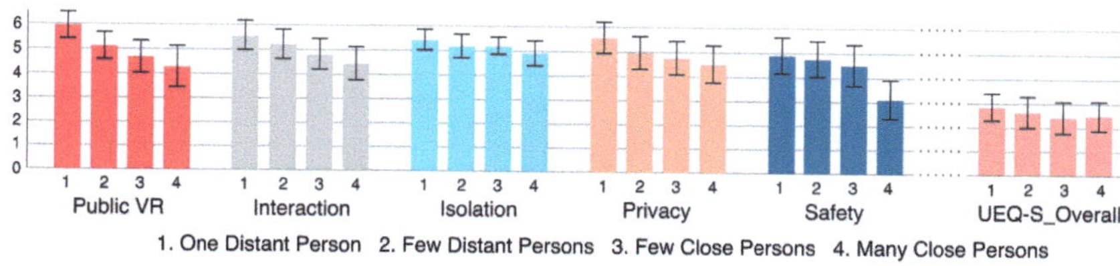

Fig. 6.1 Average values of social acceptability of Public VR, VR Interaction, VR Isolation, VR Privacy, VR Safety, and UEQ-S Overall, for Social Environments. Whiskers denote 95% confidence intervals

6.1.1.2 Technical Adoption by Social Acceptability

Analysis of data for social environments revealed that on average Public VR dimension is more acceptable when one distant person (M = 5.929, SE = 0.268) is next to the user in public settings, compared to the many nearby people (M = 4.321, SE = 0.412). When it comes to the acceptability of VR interactions, participants reported a significantly bigger acceptance of interactions for the Social Environment and for one distant person (M = 5.571, SE = 0.288) compared to interactions with many close persons (M = 4.482, SE = 0.329). The next dimension analyzed revealed that VR Isolation is significantly more acceptable for the Social Environment when again just one distant person is present (M = 5.482, SE = 0.198), as compared to when several close persons are present (M = 4.946, SE = 0.247). Furthermore, it was discovered that the privacy concerns are significantly greater for the condition with one distant person (M = 5.571, SE = 0.307) compared to the condition with many near individuals (M = 4.554, SE = 0.367) when it comes to the acceptability of VR privacy in a social environment. Participants, when asked to rate the acceptability of VR Safety, stated that they felt significantly safer in the Social Environment with one distant person (M = 4.893, SE = 0.365) than they did with many people near them (M = 3.125, SE = 0.386). These findings are shown in Fig. 6.1.

Additionally, when comparing effects based on the two types of interactions, the findings from the VR Safety dimension showed that participants reported feeling significantly safer during static interactions (M = 4.321, SE = 0.306) as opposed to dynamic ones (M = 3.982, SE = 0.295), as shown in Fig. 6.2.

6.1.1.3 Presence

Another interesting result regarding the influence of different interactivity in social environments was reported for the feeling of presence users have reported. Participants had a considerably lower average general presence value for static interaction (M = 4.134, SE = 0.130) in comparison to the dynamic one (M = 4.446, SE = 0.102), as it is shown by Fig. 6.2.

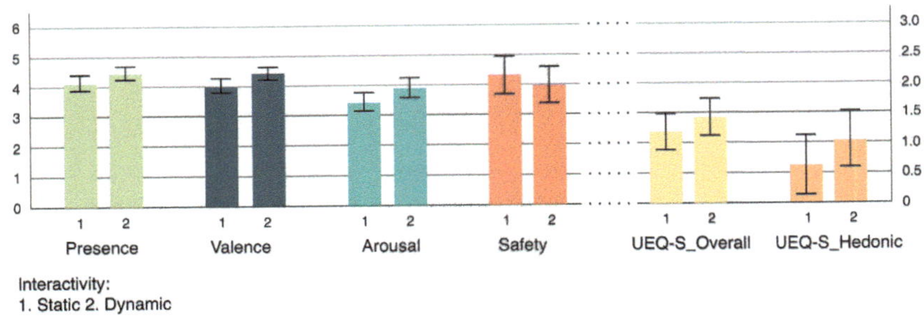

Fig. 6.2 Average values of General Presence, Valence, Arousal, social acceptability of VR Safety, UEQ-S_Overall, and UEQ-S_Hedonic quality, for the degree of interactivity in social environments. Whiskers denote 95% confidence intervals

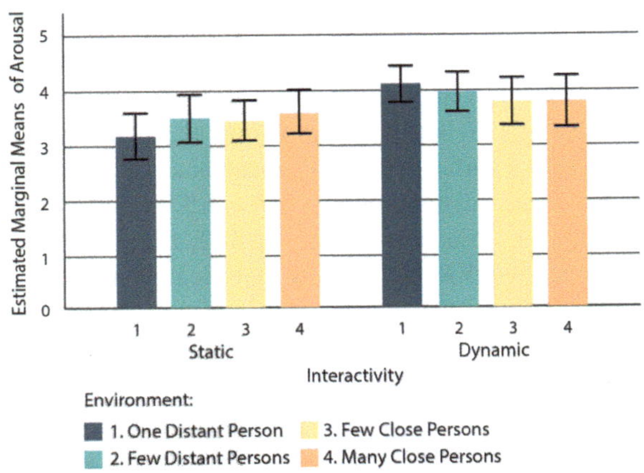

Fig. 6.3 Average values of arousal, for the combined effect of the degree of interactivity and Social Environments. Whiskers denote 95% confidence intervals

6.1.1.4 Emotions

Effects regarding emotional responses for users have mainly been associated with different interactions when in social environments. The level of interactivity was reported to significantly influence the SAM dimension valence as research results have demonstrated that static interactions, even in social settings, are associated with a lower level of pleasure (M = 4.027, SE = 0.124) compared to the dynamic one (M = 4.420, SE = 0.104). But, participants have stated that the dynamic interaction (M = 3.911, SE = 0.163) evokes a higher level of SAM Arousal value than the static (M = 3.438, SE = 0.152).

The results show that there is a statistically significant interaction between the two independent variables, Social Environments and Degree of Interactivity, which influences the SAM dimension arousal. The outcomes of combining the static and dynamic conditions resulted in opposing tendencies, as shown in Fig. 6.3. In the static scenario, the degree of arousal grows as more people enter the social environment, while for the condition with the dynamic scenario, the arousal value decreases as more people enter the social environment.

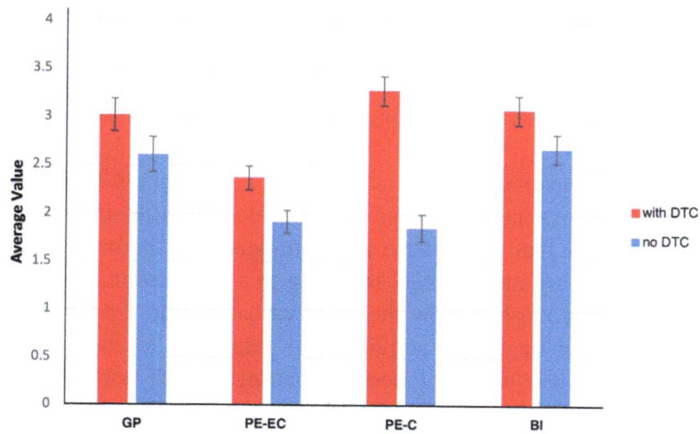

Fig. 6.4 Mean value for NMQ G1, PE-C, PE-C, and BI value for the conditions. Whiskers denote 95% confidence intervals

6.1.2 Influence of Social Conversation

6.1.2.1 Presence

The influence of conversation between users during gameplay of VR serious games was explored in Study 1, where depending on the condition, users had a possibility of direct conversation (DTC) between each other, and the influence of this conversation opportunity with an only social purpose and no additional task was explored in relationship to presence as a component of UX measured by IPQ and NMQ.

When comparing the condition with DTC, the general presence was rater significantly higher (M = 3.02, SEM = 0.17) compared to the condition where users had no option for conversation (M = 2.61, SEM = 0.18). Additionally, regarding social presence measured by NMQ, the perceived emotional contagion is significantly higher with DTC (M = 2.37, SEM = 0.12) than for the condition without DTC (M = 1.92, SEM = 0.12)). The same trend is visible as well for perceived comprehension, where the value with DTC (M = 3.28, SEM = 0.15) is significantly higher than the value without DTC (M = 1.86, SEM = 0.14), and the behavioral interdependence dimension (which refers to the mutual impact that people have on each) was significantly higher with DTC (M = 3.08, SEM = 0.15) compared to without DTC (M = 2.68, SEM = 0.15). Results are illustrated with Fig. 6.4.

6.2 Discussion

In order to find out how social environments affect UX (RQ3.1), four different levels of social environments were simulated in VR through 360° videos: one person far away, a few people far away, a few people close by, and a lot of people close by to the user. Additionally, to explore if different interactivity levels affect the perception of social acceptability, as part of this research, two different games were included, one having static gameplay not including full-body movements, while the other was dynamic.

In particular, dimensions of social acceptability (see Sect. 2.2.6) involved were those including the acceptability of public VR, interaction, isolation, privacy, and security. Results show that when there are more people in the user's surroundings, the scale of social environments rises, resulting in a decrease in the average UEQ-S score. This could be explained by the fact that having more people in the user's area suggests a higher degree of activity. This might be because of noise in the surroundings as a consequence of people chatting, which acts as a distraction. This idea is supported by another study [172], which report noise and freedom of interaction as contributing to the autonomy.

Furthermore, the social context had varying effects on the characteristics of social acceptability. The findings reveal that if it is only a matter of going from one far person to a few distant persons from the user, the acceptability of using VR in public is significantly reduced. Social acceptability reduces with each incremental addition to the number of persons around or located in close proximity to users. The reasons for this may vary, but the fact remains that consumers are concerned about using such equipment in public. A similar trend was discovered in other aspects of social acceptability. In terms of the social acceptability of the interaction, the results revealed a significant variation depending on whether the users were surrounded by a few or many people close by. Acceptability lowers significantly again when more people are nearby the user, and the qualitative results of the questionnaire revealed that users are concerned that by moving their bodies during gameplay, they would not harm themselves or the person next to them. Moreover, this result can also be related to the study [198], which pointed out people's concerns about their movements' visibility. This explanation also reasons why the lowest social acceptability levels were recorded for safety. The scale of the social environment has a significant influence on safety, and the results demonstrate that when comparing the conditions of distant people to those of close people, the acceptability of this dimension reduces significantly, owing to the increased risk of injury. It is fair to assume that for people who use VR in social contexts, safety is a continual concern. In addition, when it comes to privacy issues, they continue to be quite concerned about viewers filming them in public. These worries are understandably far more significant, and as a result, they are less tolerable in crowded circumstances. This result is not matching with the one from study [172], probably because the university setting in which their experiment was performed was creating some reason why using VR.

Moreover, regarding emotional responses, the SAM Arousal dimension is the only one where different social settings concerning the degree of involvement give significant findings. Data suggests an opposite pattern, with arousal increasing with the number of people surrounding users for static games while decreasing for dynamic gameplay. What may be claimed is that the study on the simple main effects revealed an overall trend that verifies the significance of the degree of interactivity on arousal, implying that a more specific reason for the opposing tendencies needs additional research. Despite the results, most participants stated that they would engage in VR activities if the circumstances allowed for it. On the other hand, virtual reality isolation with other persons in close proximity does not appear to be fully considered as such.

Additionally, when it comes to including the opportunity for conversation in multiplayer VR serious games (RQ3.2), results have shown that there are no significant effects on flow found but rather on the feeling of presence and social presence (see Sect. 2.2.3). According to the results, the IPQ dimension is affected by the opportunity to have conversations in VR multiplayer serious games. In particular, there has been an impact on the manner in which individuals sense the general presence. When the opportunity for dialogue was made available, the general presence values were much higher. However, when comparing values with conversation possibility, with or without a virtual environment, results report that those ones using VR with conversation have the highest reported feeling of general presence. Nevertheless, the major impact of adding conversation opportunities to the multiplayer game resulted in significantly different levels of social presence, which is also supported by theory of social presence [143]. According to the results, the circumstance in which there was the opportunity for dialogue significantly impacted the NMQ subdimensions measuring the social presence. Even though communication between players was not mainly happening during the race but before or after, it still led to a strong social effect. In particular, comprehension could be explained by the fact that users were aware of their opponents' situation and had the chance to congratulate or comment on the race, which led to better emotional understanding. Overall, the findings demonstrated that incorporating a competitive multiplayer exergame into a virtual setting and providing a means for the players to communicate with their opponents were highly rated by users.

Altogether, when it comes to influences of social context on UX of VR serious games, these studies have found several influences that are shown in Fig. 6.5. As two different studies were focused on different factors, namely different social environments or conversation opportunities, results have reported that both have a significant effect on the reported feeling of presence and social presence. Additionally, different social environments have influenced also users reported emotional responses and overall experience, as well as a majority of dimensions of social acceptability. As the social environment becomes more crowded, it produces movement restrictions and embarrassment and raises privacy and security issues.

6.2.1 Content Design Guidelines

Based on the findings and conclusions of user research, the following guidelines are established, with a focus on the social environment of VR serious games:

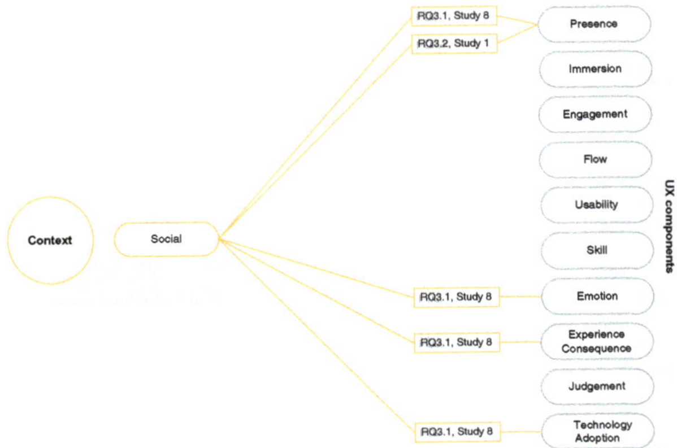

Fig. 6.5 Illustration of connections between Context IF, particularly Social, to UX components where significant effects from a particular study contribute to answering RQs

- Users are always concerned about their safety and privacy. With more people nearby, the acceptability of using VR drops significantly because users are worried that moving their bodies during games would cause them to injure themselves or the person next to them or that someone will record them.
- Even in a social situation, dynamic interactivity will improve feelings of presence, engagement, pleasure, and excitement and so may be an additional motivation to use VR solutions. However, a safe social environment should be provided.
- Users enjoy the inclusion of a virtual environment with the ability to communicate with one another. To increase the sense of social presence, the option to communicate with another user in a VR serious game should be provided.

6.3 Summary

This chapter investigated VR serious gaming scenarios and user evaluations. Because of the limits of lab-based research and ethical user studies, most context subfactors were treated as controlled variables, while the emphasis for the studies was put on the social influencing factors. The goal was to investigate how social context influences VR's social acceptability and users' emotional reactions (RQ3.1), including how communication influences users' experiences (RQ3.2), with Studies 1 and 8.

Several different components of the UX model have been influenced by different social environments. Using 360-degree videos, a simulation of four distinct social situations, and two VR games with varying degrees of player interaction, was produced. The data revealed a statistically significant relationship between social environments and overall UX, public VR, interaction, isolation, privacy, and safety acceptance. The results also revealed a statistically significant impact of interaction level on presence, valence, arousal, overall UX, UX hedonic quality, and safety acceptance.

The impact of the social context, which had a varied effect on practically all characteristics of social acceptability, is especially interesting. The findings reveal that if it is simply a question of shifting from one far person to a few distant persons from the user, the acceptability of using VR in public is substantially diminished. Social acceptability lowers with each new rise in the number of persons surrounding or placed close to the user.

Furthermore, a player's social presence may also be increased by providing them with the opportunity to interact with other players. The findings that were given demonstrated that the participants had a positive reaction to the introduction of a multiplayer exergame for a virtual environment that included the potential of communication between players competing against one another.

In conclusion, a summary of all the outcomes can be viewed in Fig. 6.5, which shows how Context Social IF affects UX components.

Chapter 7
Human IF

7.1 Results

This chapter focuses on users engaged in a VR serious game and how various user attributes might impact UX. Because this study relies on questionnaires for assessment, physiological measures were excluded. As a result, demographic questionnaires were employed in all studies. Furthermore, immersion tendency and simulator sickness were not assessed for all conditions but were regulated throughout the tests. Age and gender (RQ4.1) were particularly highlighted as demographic factors, as was a prior experience with VR (RQ4.2). This study examined how users of various ages, genders, and levels of experience perceive the user experience in VR serious games.

Study 2 and Study 7 were two user studies conducted using Human IF to investigate these effects (where Table C.2 provides a summary of details of these studies). A repeated measure Analysis of Variance (ANOVA) was used to detect statistically significant differences in the results. Tables 7.1 and 7.2 provide an overview of all significant effects that will be discussed next.

7.1.1 Influence of Demographics by Age and Gender

In order to investigate the impact of age and gender as demographic factors, participants in Study 2 and Study 7 were divided into age and gender groups, and UX component measures such as UEQ for an overall score of UX, FSS for the feeling of flow, and questions about support, clearness, and level of concentration or distraction were used, revealing significant differences in results.

T. Kojić, *User Experience for Serious Games in Virtual Reality*, T-Labs Series in Telecommunication Services, https://doi.org/10.1007/978-3-031-75530-9_7

Table 7.1 The effects of varying UI complexity and positioning, in addition to covariations of user previous experience with VR (VR_x), user gender (gndr), and user age (age) on UEQ pragmatic quality (UEQ_prag) and UEQ overall quality (UEQ_all); on the perception of flow (flow); and on clearness (clear) and support to the user during exergame (support)

Parameter	Effect	df_n	df_d	F	p	η_G^2
Complexity and VR_x	UEQ_all	1	21	4.77	0.040	0.19
Complexity and gndr	UEQ_all	1	21	7, 35	0.013	0.26
Complexity and gndr	UEQ_prag	1	21	7.16	0.014	0.25
Complexity and age	Support	1	21	11.49	0.003	0.35
Complexity and age	Clear	1	21	8.95	0.007	0.30
Position and gndr	Flow	1	21	6.62	0.018	0.24
Position and age	UEQ_all	1	21	5.00	0.036	0.19

Table 7.2 Effects of system version (2D paper vs. 3D VR) in addition with covariations of user previous experience with VR (VR_x)) and user age (age) on distraction (DIST), concentration (CONCT), and recognized patterns (PAT)

Parameter	Effect	df_n	df_d	F	p	η_G^2
2D/3D and age	DIST	1	23	5.230	0.036	0.246
2D/3D and age	CONCT	1	23	4.675	0.044	0.229
2D/3D and VR_x	PAT	1	23	6.260	0.024	0.281

7.1.1.1 Experience Consequence by UX

The findings of Study 2 revealed that age covariance along with positioning had a significant influence on UEQ total score. Older participants evaluated the overall UX of representation on a boat higher (M = 1.18, SD = 0.76) than cockpit representation (M = 1.01, SD = 0.81), whereas younger people rated higher the overall UX of the cockpit (M = 1.61, SD = 0.69) than metrics on a boat (M = 1.35, SD = 0.88). In correlation with gender, a similar pattern of opposing preferences is reported. Women gave a better overall UX score for digital (M = 1.56, SD = 0.95) than the gamified version (M = 1.23, SD = 1.07), whereas men gave a higher overall UX score for the gamified version (M = 1.27, SD = 0.57) than the digital (M = 1.07, SD = 0.69). Gender also has an impact on the UEQ pragmatic quality. Whereas women preferred pragmatic quality for digital (M = 1.89, SD = 0.84) above gamified metrics (M = 1.44, SD = 1.04), men preferred the reverse.

7.1.1.2 Usability by Support and Clearness

Different preferences were discovered in Study 2 in terms of perceived amount of support and clarity of scenes in correlation with age. When asked how helpful metrics representation was to them, younger participants assessed gamified visualization of metrics as more supporting (M = 6.32, SD = 0.78) than digital visualization (M = 5.95, SD = 0.72). This preference was reversed for older indi-

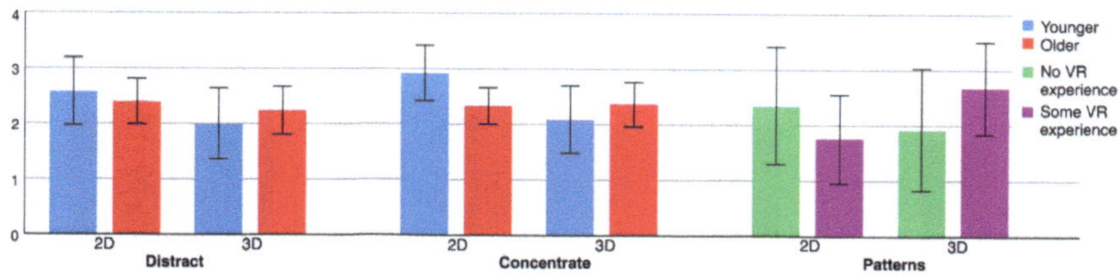

Fig. 7.1 Average subjective assessment across all participants for 2D paper and 3D VR versions (from left to right) on degree of distraction and concentration based on participant age group and pattern recognition based on the amount of prior experience with VR. Whiskers denote 95% confidence intervals

viduals, who preferred digital visualization (M = 6.13, SD = 0.97) over gamified visualization (M = 5.7, SD = 0.92). Similarly, younger participants thought the gamified version was clearer (M = 6.68, SD = 0.48) than the digital (M = 6.05, SD = 0.79), while older participants thought the digital was better (M = 6.47, SD = 0.82) than the gamified (M = 6.37, SD = 0.85).

7.1.1.3 Flow

Another influence in which gender had a significant role was the experience of being in flow (Fig. 7.2b). Women reported feeling more in flow while using metrics representation as following boat (M = 4.34, SD = 0.39) compared to cockpit metrics (M = 4.14, SD = 0.53), whereas men reported feeling more in flow with cockpit (M = 4.19, SD = 0.29) compared to boat representation (M = 4.06, SD = 0.46).

7.1.1.4 Usability by Distraction and Concentration

Moreover, Study 7 data show that age is a covariate when it comes to the reported level of concentration (Fig. 7.1). While the younger group reported being more concentrated using paper (M = 2.56, SD = 0.38) and significantly less distracted (M = 2.57, SD = 0.53) compared to concentration (M = 2.14, SD = 0.38) and distraction in VR setting (M = 2.00, SD = 0.57), the older group reported being opposite with less concentration (M = 2.31, SD = 0.60) and more distracting (M = 2.44, SD = 0.73) for paper and more concentration (M = 2.37, SD = 0.72) and less distracting (M = 2.24, SD = 0.77) for VR.

7.1.2 Influence of Expectations and Expertise

7.1.2.1 Experience Consequence by UX

Concerning the influence of user expectations and expertise on UEQ overall quality, significant variations in covariance with user experience in using VR systems

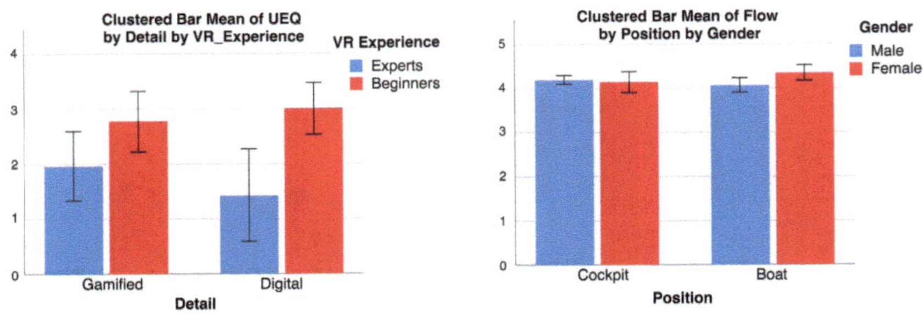

(a) UEQ (b) Flow

Fig. 7.2 (**a**) Mean UEQ overall (UEQ_all) value of all participants for the conditions with gamified and digital visualization split by VR expertise of participants. (**b**) Mean flow value of all participants for the conditions with the position of visualization as the cockpit and in a boat split by gender. Whiskers denote 95% confidence intervals

have been found (Fig. 7.2a). In Study 2, not-so-experienced users rated the digital version for metrics for overall UX (M = 1.5, SD = 0.71) higher than the gamified version (M = 1.38, SD = 0.83). However, users with more VR experience evaluated gamified visualization's UX higher (M = 0.98, SD = 0.63) than digital (M = 0.71, SD = 0.84).

7.1.2.2 Usability by Recognized Patterns

In Study 7, the results also show significant differences in preferences when participants' prior experience with VR is taken into account (Fig. 7.1). This effect was related to pattern observation, with participants with no VR experience finding more familiar patterns for paper (M = 2.29, SD = 1.11) compared to VR (M = 1.86, SD = 0.69), but those with some experience finding more familiar patterns in VR (M = 2.50, SD = 1.15) compared to paper (M = 2.06, SD = 1.06).

7.2 Discussion

When it comes to the configuration of VR serious games, the results have suggested that different people with different levels of prior expertise with VR systems, genders, and ages will have varied preferences. In particular, for exergaming, it resulted in diverse results of user experience, clearness of UI, perceived support, and a sense of flow. Further on, for the financial data exploration, users reported a different degree of concentration and distraction, as well as different recognized patterns.

The "wow effect" that a person has while using VR for the first time is generally characterized as a sense of surprise or excitement, and this influence eventually diminishes as the user grows more used to using VR [209]. However, this effect might be observed not just in VR but also in other sectors like gaming, cinema, and even in storytelling. According to VR exergaming study results, participants with more previous expertise with VR systems evaluated better overall quality for the gamified visuals. This is most likely because these individuals are already used to viewing gamified virtual surroundings, while others who encountered VR for the first time might have been overwhelmed by either the device or the environment. Similarly, when it comes to the visualization of financial data in VR, different viewpoints on the capacity to determine patterns have been reported. Individuals with no previous experience with VR reported seeing more patterns in the paper version, whereas those with some prior acquaintance saw more patterns in VR, which again could be explained by the fact that they were used to viewing virtual surroundings, and such patterns were also known to them. The expertise of players in gaming has been shown to be an influencing factor when rating video games [186].

Age is another important factor that results have reported having a role in users' preferences. When rating how easy it was for participants to concentrate on a task with financial visualizations, the results have shown that, interestingly, older age groups reported that it was easier to concentrate in VR. However, those users are not old but just placed in the older group for this study. Furthermore, when considering the player's age in VR exergaming, the overall quality of the visualization of metrics on a boat was assessed higher by older participants, while the overall quality of the cockpit visualization was rated higher by younger people. A similar trend was discovered with the difficulty of the task at hand as the age of the participants affected the clarity of the UI. Younger participants felt that gamified visualization was easier to understand, whereas those who were older felt that digital visualization was easier to understand. Such effects of age were expected as age was proven to be a factor already for computer experiences [204].

However, gender also impacted the overall UX rating, with men giving a higher rating to the overall UX of a gamified visualization and women giving a higher rating to the overall UX of a digital visualization. There was also a gender difference in the pragmatic quality results, with the same trend seen in the pragmatic quality of gamified visualization of metrics assessed by men, yet women rated the pragmatic quality of digital visualization higher. According to the results, the participants' impressions of the flow condition differed depending on where the measurements were placed and their gender. The flow experience was raised for men in situations where metrics visualization was displayed on a cockpit, whereas the sensation of flow was increased for females in situations where metrics visualization was placed in a boat. When comparing just the different UI conditions, however, the study found no significant differences in the flow state of the users. This might be because none of the conditions were interfering with their activity or distracting them from it, which was one of the design goals supported by user-centric design [13]. Another design aim was to provide data while still providing an immersive experience in a virtual world, and differences because of gender, similar to age, can be explained that VR experience is a version of a computer experience [204].

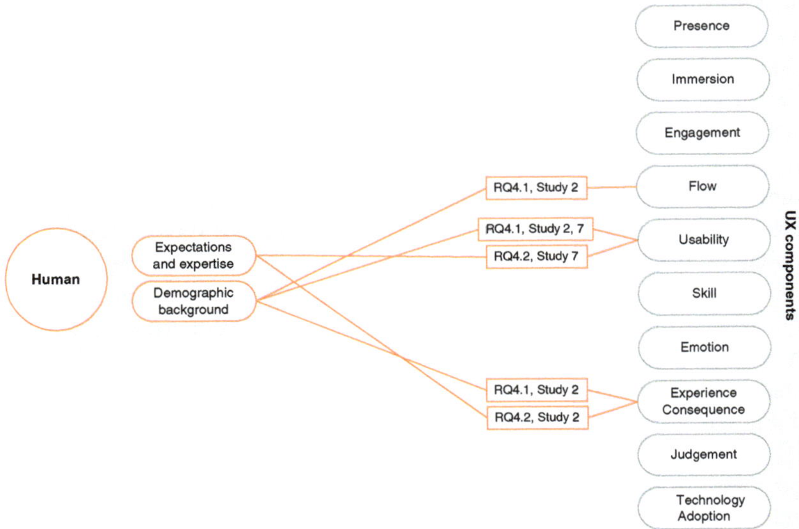

Fig. 7.3 Illustration of connections between Human IF, particularly Expectations and Expertise and Demographic Background, and UX components where significant effects from a particular study contribute to answering RQs

This demonstrates that the user's history, including demographic information such as age and gender and previous experience with VR systems, is significant and needs to be considered when designing a user interface for VR serious games to provide users with the best possible experience. When it comes to how human factors influence the UX of VR serious games, user studies done as part of this research have shown that subfactors such as Expectations and Expertise, as well as Demographic Background, significantly alter user preferences, as illustrated in Fig. 7.3. For Usability as a UX component, both studies have reported significant differences in both influencing subfactors. Both subfactors, including different ages, gender, and previous VR experiences, have also shown that they influence preferences regarding the overall UX score included in the Experience Consequence from the UX model. However, regarding the different feelings of flow, the influence of the results was reported based on different demographic backgrounds.

7.2.1 Human Design Guidelines

Based on the findings and conclusions of user research, the following guidelines are established, with a focus on the social environment of VR serious games:

- In order to provide the best possible experience, it is necessary to consider the user's attributes, which may include demographic information like as age,

gender, and previous experience with VR systems while developing VR serious games.
- Age, in addition to gender, had a role in user preferences and had an influence on the overall UX rating. Design based on preferences for the majority of users but allow changes in settings.

7.3 Summary

This chapter explored user preferences about various motivations for using VR serious games, with the goal of determining if users have distinct preferences according to their individual user characteristics. Age, gender, and differing prior VR experience of users representing subfactors of Expectations and Competence and Demographic Background have been placed in the focus of IFs. The goal was to discover how users of various ages and genders perceive user experience (RQ4.1) and how past VR experience influences user preferences in VR serious games (RQ4.2). As a result, two experiments were undertaken with different serious games in mind, with Study 2 representing an active serious game focusing on exergaming and Study 7 focusing on personal financial data research.

When it comes to the ways in which human factors affect the user experience of VR serious games, studies revealed that both subfactors Expectations and Expertise and Demographic Background significantly alter user preference. According to study findings, when it comes to the setup of VR serious games, different people with different extents of previous expertise with VR systems, genders, and ages would have different preferences; in particular, for VR exergaming resulted in a variety of user experiences, UI clarity, perceived support, and a feeling of flow. Furthermore, individuals reported varying degrees of focus and distraction, as well as distinct recognized patterns, while exploring financial data.

Results have shown that depending on age, preferences on whether it is easier to focus in or outside VR will vary. Similarly, participants judged the overall quality of the visualization of metrics differently depending on their age, with older individuals selecting a realistic picture of a boat, while younger people reported that a gamified visualization was simpler to comprehend. Similarly, results show differing preferences depending on gender, with men rating the overall UX of a gamified visualization higher and women giving the better score for the overall UX of a digital visualization. Also, individuals with greater previous familiarity with VR systems gave higher ratings to the overall UX for the gamified visuals. Finally, users who had never used VR before reported detecting more patterns in the paper version that they were more used to while reviewing financial data. Because of those results, the user's demographic information, such as age, gender, and previous VR experience, must be considered while building serious VR games; as such an approach can provide the greatest possible experience for users.

The results are summarized in Fig. 7.3, illustrating how studied Human IFs (age, gender, and previous VR experience) impact UX components.

Chapter 8
Conclusion and Outlook

This thesis is focused on creating an overview between User Experience and Influencing Factors for VR serious games. The basis for this work was the UXIVE model composed of ten components extracted from existing models (presence, engagement, immersion, flow, usability, skill, emotion, experience consequence, judgment, and technology adoption), the ITU-T Rec. G.1032 [188] that focused on influencing factors of gaming quality of experience, and the ITU-T Rec. G.1035 [15] that categorizes and analyzes the influencing factors for VR service.

VR serious games are a special category that partly fit in mentioned categories of VR service and gaming but are special because their primary purpose is not entertainment, but they have some other goal (Sect. 2.1.3). In VR serious games, the concept of gamification, which came from the gaming industry, is used as an opportunity to make use of the intrinsic motivation that games provide [192]. Motivation and engagement are also important aspects of UX; hence the idea behind using VR is to enhance UX by making people more self-motivated via gamification and because users in such cases can enjoy experiences that they cannot have in real life and all of this by giving them the ability to control and think in virtual environments [130]. This idea, in particular, is building on the Self-Determination Theory (SDT) [130] as a psychological concept that describes how individuals live and how they select what they want to do (Sect. 2.2.2). Furthermore, VR serious games do share specifics of VR services such as the feeling of presence, immersion, simulator sickness, and social acceptability (as mentioned in Chap. 2), which are important concepts for UX as well.

Therefore, when discussing UX for VR serious games, work and results were divided into four categories: System (Chap. 4), Content (Chap. 5), Context (Chap. 6), and Human (Chap. 7). This was done as recommendations for IFs of VR services are distinguishing only between three main categories, but gaming field also recognizes Content as important. However, not all IFs known from VR services and gaming were explored in this thesis, but there were limitations (see Sect. 2.3.3) due to the methodology (user studies in the laboratory) and due to VR devices

T. Kojić, *User Experience for Serious Games in Virtual Reality*, T-Labs Series in Telecommunication Services, https://doi.org/10.1007/978-3-031-75530-9_8

used (on market HMD devices). Still, for every IFs that were explored in this work, no matter if originally coming from VR services overview or gaming overview, significant results were found.

Regarding System IFs, both influences of hardware and network transmission turned out to be important factors, each influencing several UX components. These results go in line with the current ITU-T Rec. G.1032 but are pointing out several additional insights as well. Hardware IFs so far in recommendation have been considered from a technical point of view and how improvements in weight, size, resolution, and field of view (as in Sect. 2.3.2.1). However, the results presented here report that with the use of a market-ready HMD device, the presence, involvement, and experienced realism (as IPQ measurements shown in Sect. 4.1.1) were rated significantly higher. This is finally replicated in results that show higher flow and motivation for conditions using VR for serious games as well. These results align with SDT and motivation theory, which has been presented in the gaming field and recognized as a factor.

Content IFs have not been subcategorized in the current ITU-T Rec. G.1032, but subsections have been introduced again from the gaming fields. In particular, for this work, examined subfactors were Supported Interactions and Environmental Design. Similar to System IFs, all of them have shown to be also influencing factors for VR serious games, resulting in statistically significant results on UX components. During the design of VR serious applications, the current guidelines regarding UI design and development were followed, such as Google guidelines and WCAG2.1 (explained in Sects. 2.1.2.4 and 2.2.5). However, when it comes to the results WCAG2.1, it turned out to be not a suitable standard for the display of text in VR serious games, as participants in the study have reported being the worst setting for users while using the best level AAA contrast. Still, the results of this study are not proposing new values for the standard of contrast values but propose to include Context IFs (particularly Supported Interactions and Environmental Design) in the overview for IFs of VR serious games.

When it comes to Context IFs, due to limitations in methodology, only Social IF was included in this work and has shown some very interesting results. In the current overview ITU-T Rec. G.1035 includes Social Context as IF, but it is mentioned that this factor requires further investigation and is covered on a hypothetical level. Limited research regarding Social Context seems to be the case for VR in general (see Sect. 2.2.6), but results from this work present that Social Context seems to be an important factor. Users have reported their concerns about their safety and privacy, which is not the case or reported in the gaming field. As more people are nearby, the acceptability of using VR drops significantly. Although done in a controlled environment, this study is the first step and is already showing the importance of including Social IFs in the overview.

Also, for Human IFs, two investigated factors were Expectations and Expertise and Demographic Background. Similar to gaming previous experience and demographics of players in gaming, these aspects were investigated for VR serious games. Building on top of the UXIVE model, results have shown again that both are influencing more than one component of UX and additionally have also reported different preferences of users depending on their demographics. In the current overview of ITU-T Rec. G.1035, demographics are included as part of the Human IFs, but only because changes in ITU work have already been made based on the results of this work.

In summary, the main contributions of the thesis are design guidelines based on the user-centric design principles for creating VR serious games and based on the empirical validation of influencing factors for VR serious games, together with the insights of how influencing factors are connected with the particular aspects and components of UX.

8.1 Answers to Research Questions

This chapter will address the responses to the research questions given at the beginning of this dissertation. The research questions will demonstrate how the thesis's main goal of evaluating user experience for VR serious games was addressed.

RQ1.1: Can the use of a virtual reality set-up increase the users' motivation and engagement for VR serious games?

The introduction of a virtual environment resulted in an increase in the feelings of general presence, spatial presence, involvement, and perceived realism, according to the results. These results suggest that users could find their virtual experiences genuine and engaging, even more than some of the tasks represented in the real world. Additionally, the participants' subjective views in VR significantly differed when the feeling of time was compared to the system's real progress time.

Forgetting about the time while gaming or reporting on the difference in passing the time is an influence that happens when users are in the flow of activity, and in this case, it may have happened because the VR solution provided users with more than just an activity to complete. Because VR solutions seem to enhance the sense of presence and put users more in the zone, one should consider using them for serious purposes as well because users with a strong motivation will want to finish the assignments.

RQ1.2 Does the network delay during VR serious games have an influence on the user experience, in particular regarding the perception of own delay?

The sensations of presence and flow as UX characteristics were investigated to determine whether or not delay influenced the experience of playing VR serious games, and participants were explicitly asked whether they observed any delays during the experiment. Results report that there is less of a sensation of presence and flow when there is more delay. These findings, which appeared as visual jitter and image skipping due to network delays, may also be correlated with video game findings. Compared to real life, where the world never stops moving in front of a person, this kind of fragmented visual display seems strange to users.

It is also worth mentioning that users could not determine if the network problem was on their side or their opponent's. According to the results, a player experiencing a network delay cannot properly assess the situation and incorrectly assumes that their opponent is also delayed, which means that when a player has a bad network, he feels that the quality is poor, not just because of own settings but also because of the opponent's network.

RQ2.1 Does interaction ability and generated content in a virtual environment have an impact on the user experience in VR serious games, in particular on the perception of flow and presence of players?

Studies assessing the influence of content and virtual environments reported results that revealed how different types of activities changed users' preferences for the VR serious game. The different interactions presented influenced all three SAM dimensions (valence, arousal, and dominance). The results show that participants felt less excited when using hand tracking interactions in virtual environments. However, the fact that users like the experience may contribute to a higher valence for hand tracking. Still, a lower feeling of dominance may emerge from the fact that engaging in hand tracking results in a lower sense of control or precision. Overall, this implies that the virtual environment affected the user's emotional response while engaging in the activity.

In addition, the influence of the virtual environment was evaluated for the user interface design, specifically text, and participants were given a task in which they were asked to identify the setting in which they felt most comfortable reading. The results suggest that the length of the text and the device on which it is shown significantly influence how consumers perceive the experience of reading in VR. Even though VR provides more space, participants did not like bigger font sizes. They intended to make a big angular size choice for the negative option, and they picked a big font for the text that was shown at a very small distance. As a consequence, the average values for negative readability instances were much higher than those for positive reading examples. Such cases are not yet specified in the current standards that offer only minimal requirements for font sizes.

RQ2.2 How do the differences in complexity and position of user interface elements influence user experience in VR serious games?

Different representations of metrics from a VR rowing training game were displayed to investigate the influence of UI design on UX and the manner in which elements may be positioned inside a virtual environment with relation to the placement and complexity of these components. Using the data presented, it has been shown that the position and complexity of UI, such as metrics in rowing, have an influence on the experience that one gets while participating in VR exergaming. Overall, participants rated the feeling of support and legibility differently for different graphics. The cockpit with gamified visualization and the boat with digital presentation of the measurements were picked as the top two favorite visualizations by the participants overall. However, both of those metric representations seem to have advantages and disadvantages in comparison to one another, and the eventual design decision must be suited to the user or flexible enough to enable the possibility of several configurations depending on the profile of a certain user.

RQ3.1 Does the social environment influence user experience, resulting in the social acceptability of VR serious games?
According to the results, diverse social contexts influence the feelings and sense of presence experienced by users, as well as their overall impressions of the UX and almost all aspects of social acceptability. The results indicate that the acceptability of VR technology in public places is significantly reduced even with only shifting from one person to a few distant persons from the user. The number of people who are nearby or who are positioned close to users has a direct impact on the level of social acceptance that they enjoy. There could be many different reasons for this, but the results report that users feel uncomfortable using this kind of technology in public. The same pattern was found in the investigation of several factors of societal acceptability. It is acceptable to hypothesize that individuals who use VR in social settings constantly worry about their safety. In addition, when it comes to matters related to their privacy, they continue to be highly worried about spectators recording them when they are out in public. These concerns are, for understandable reasons, a lot more relevant and harder to regulate when the environment is crowded.

RQ3.2 How does the opportunity to have conversations in multiplayer VR serious games influence social presence?
Regarding whether or not it is beneficial to include the option for conversation in multiplayer VR serious games, the findings indicate that this decision does not have a significant impact on flow but rather on the sensation of presence and the sense of social presence. The general presence ratings increased once it was made possible for participants to engage in conversation. Nevertheless, when values are compared with the option of having a conversation, the findings show that people who used VR with conversation had the greatest experience of general presence in the experiment. According to the findings, the situation where there was the potential for conversation significantly influenced the subdimensions assessing social presence. These subdimensions include perceived emotional contagion, perceived understanding, and behavioral dependency.

RQ4.1 How do age and gender affect the perceived quality and user experience for VR serious games?
Both age and gender have been shown to influence aspects of UX for VR serious gaming. The study findings based on VR rowing exergaming revealed that the participants' perceptions of the flow varied with regard to the participant's gender. The impression of flow was enhanced for women when metrics visualization was put in a boat, whereas the flow experience was enhanced for men in situations in which metrics visualization was exhibited on a cockpit. There was also a gender difference in the evaluation of the overall UX, with men giving a higher rating to the overall UX of a gamified visualization, while women gave a higher rating to

the overall UX of digital visualization. When taking into account the age of the player in VR exergaming, it was found that older participants considered the overall quality of the visualization of metrics on a boat as better, but younger individuals rated the overall quality of the cockpit visualization as preferred. Another pattern that was identified was that the age of the participants affected the perceived clarity of the user interface. Additionally, when assessing how easy it was for participants to focus on a task using financial visualizations, the findings showed that older age groups claimed it was easier to concentrate in VR.

RQ4.2 Does the previous experience with VR have an influence on the current user experience within VR serious games?

According to the findings, individuals with various amounts of previous experience with VR systems are likely to have varying preferences regarding the set-up of VR serious games. This is because different people have different degrees of prior knowledge of VR systems. The VR exergaming study participants with some prior experience with VR systems rated the gamified solutions with higher overall quality scores. This is most likely because these people are already used to watching gamified virtual environments, unlike others who may have been overwhelmed by their first experience with VR. Similarly, for financial overview in VR, those with no prior experience with VR reported seeing more patterns in the paper version, but individuals with some prior familiarity reported seeing more patterns in VR.

Finally, the overview of all influences to components of the UX model is given in Fig. 8.1.

8.2 Limitations and Future Work

The choice of methodology resulted in certain restrictions related to the context in which the studies were carried out, given that they were conducted in a laboratory. Moreover, the participants were informed and aware that they were taking part in an experiment, and after each condition, they filled out questionnaires, which they would not do in the real world while using VR solutions. All of this could have potentially altered their perspective and responses. Also, due to the system's complexity, one of the disadvantages is that users were always asked about their experience solely through questionnaires; hence no psychological measurements were used. This work gives a first overview of the influence of social context while using VR technology. Nevertheless, exploring that dimension is limited because only VR applications have been considered. AR and MR applications might be even more crucial for social context study, allowing to study a wide range of interactions and additional content. Future research on social context would support the recognition of relevant factors and aspects for public usage of XR applications. Specifically, studying public scenarios (outdoor or indoor) is believed to be extremely relevant to improving the user experience using XR technology. It would also be interesting to understand how the presence of other people can affect

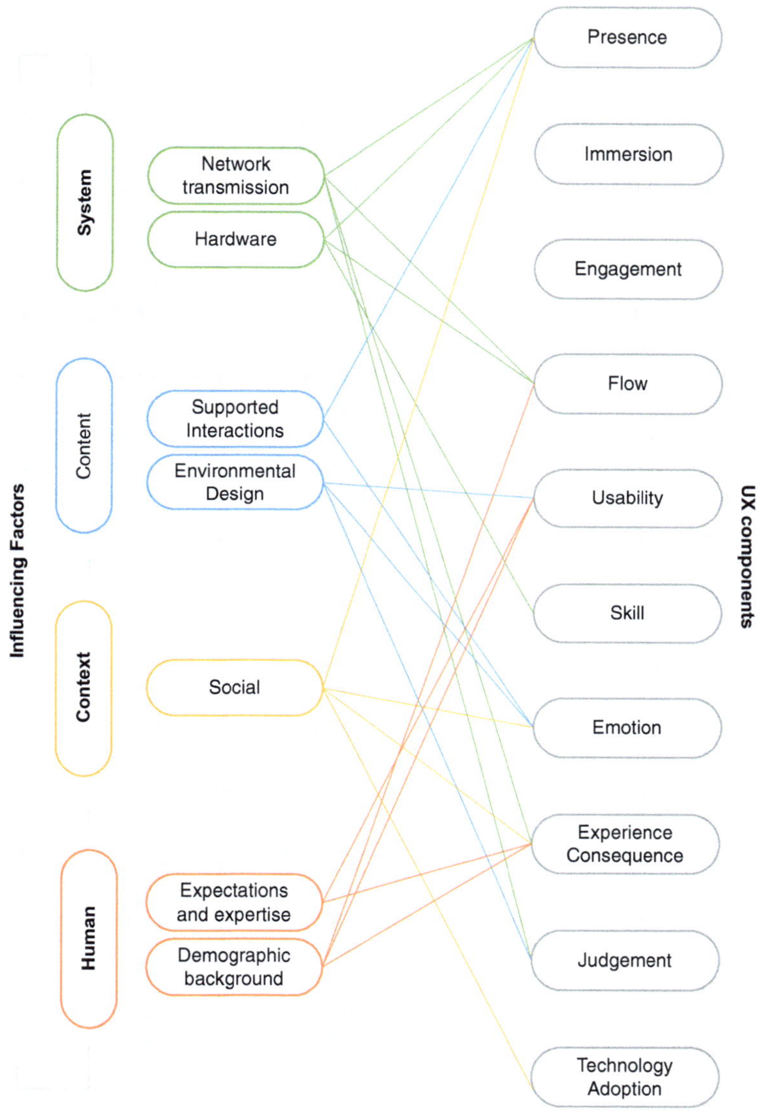

Fig. 8.1 Illustration of connections between influencing factors and particular subfactors to UX components where significant effects from a particular study contribute to answering RQs

the users' social context and how AR virtual characters blend in with the real-world experience, for example, to what extent they can have the same effect as real people and what are the users' motivations behind their preferences.

Second, the devices used were HMDs, which are currently available on the market as VR solutions. This means that no hardware features were altered on such devices and were used as it is. For future research, it would have been interesting to continue research based on new devices, as VR market is growing rapidly, and to perform and confirm effects with some of the devices that have possibilities of using not only one hardware setting. Again, here interesting future work goes also in the direction of XR solutions, possibly combining current AR and VR devices in one solution. This research also did not cover VR cloud solutions in the sense of integrating the cloud for VR games or VR serious games. As a result, communication across the network was restricted in circumstances where multiplayer was explored, and only two users were involved. Therefore, it would be interesting to explore not only VR gaming but also, in particular, VR cloud gaming. However, while doing so, it would be important to keep in mind the very important factor of simulation sickness because of sensitive network conditions. Besides that, the limiting aspect is that if at least one effect was discovered in one study, it was not confirmed with another exemplary study, but such an answer was included as soon as there was a connection between IF and a component of the UX model. So, in order to make the results more generalized, it would be necessary to repeat investigations to confirm the results based on more effects across not only one field for VR serious games.

When it comes to the generalizability of findings and users, it is important to mention that users do not ideally replicate the population of users for VR serious games. This is due to the fact that participants of the studies were recruited via the portal of Technische Universität Berlin, and even though it is not necessary to be a student to use this portal and participate in the study, still the majority of the people were students. Furthermore, particular tendencies that are part of human factors, such as immersion tendency or simulation sickness tendency, were not addressed. However, all users were informed that if the VR system made them feel bad, they could stop the experiment and still be counted as fully participating. However, no participant stopped the experiment or additionally reported any problems. Still, in the future, aspects of simulator sickness can be included, at least in a form directly asking for feedback via standardized questionnaires.

Also, when it comes to data gathering, with the best efforts put into gathering as many participants for the user studies as possible, the generalizability of data would have been better with more data; in such a way, the validity of results would have been improved.

Next, about the generalizability of the games it is once again necessary to mention that all VR serious games were applications developed as part of this work for user testing of particular aspects. This approach has enabled the possibility of controlling all aspects of test design. Nevertheless, no matter that all design principles were followed, those games are not market-ready solutions. Certainly, the quality of gameplay could be enhanced, first by improving the storyline and making the gameplay of each game longer. Therefore, future research could not only use market HMD devices but also VR serious game applications from the market.

The system's generalizability is also important to mention, even though only market-available HMD devices were used in this work. As stated in related work, VR technology offers quite a few opportunities, and when investigating VR serious games, it would be interesting to include different types of hardware solutions and set-ups, which includes not only switching between HMD and CAVE but also including different tracking devices and sensors that could enable different audio and haptic options, with the goal of expanding future work into the AR and XR field.

Overall, for future research regarding UX for VR serious games, it will be important to still keep up with the development of VR technology itself and to continue work on assessing factors that those new changes are affecting. In particular, one of the most popular VR serious games is expected to be in fields where performance will be very good, despite the fact that they are not used very frequently nowadays, merging VR technology with cloud solutions and incorporating artificial intelligence opportunities.

Appendix A
Standardized Questionnaires

AttrakDiff

Please provide your impressions of the product you have tested by check marking your impression on the scale between the terms offered in each line.

- Human – 1 – 2 – 3 – 4 – 5 – 6 – 7 – Technical
- Isolating – 1 – 2 – 3 – 4 – 5 – 6 – 7 – Connective
- Pleasant – 1 – 2 – 3 – 4 – 5 – 6 – 7 – Unpleasant
- Inventive – 1 – 2 – 3 – 4 – 5 – 6 – 7 – Conventional
- Simple – 1 – 2 – 3 – 4 – 5 – 6 – 7 – Complicated
- Professional – 1 – 2 – 3 – 4 – 5 – 6 – 7 – Unprofessional
- Ugly – 1 – 2 – 3 – 4 – 5 – 6 – 7 – Attractive
- Practical – 1 – 2 – 3 – 4 – 5 – 6 – 7 – Impractical
- Likeable – 1 – 2 – 3 – 4 – 5 – 6 – 7 – Disagreeable
- Cumbersome – 1 – 2 – 3 – 4 – 5 – 6 – 7 – Straightforward
- Stylish – 1 – 2 – 3 – 4 – 5 – 6 – 7 – Tacky
- Predictable – 1 – 2 – 3 – 4 – 5 – 6 – 7 – Unpredictable
- Cheap – 1 – 2 – 3 – 4 – 5 – 6 – 7 – Premium
- Alienating – 1 – 2 – 3 – 4 – 5 – 6 – 7 – Integrating
- Brings me closer to people – 1 – 2 – 3 – 4 – 5 – 6 – 7 – Separates me from people
- Unpresentable – 1 – 2 – 3 – 4 – 5 – 6 – 7 – Presentable
- Rejecting – 1 – 2 – 3 – 4 – 5 – 6 – 7 – Inviting
- Unimaginative – 1 – 2 – 3 – 4 – 5 – 6 – 7 – Creative
- Good – 1 – 2 – 3 – 4 – 5 – 6 – 7 – Bad
- Confusing – 1 – 2 – 3 – 4 – 5 – 6 – 7 – Clearly structured
- Repelling – 1 – 2 – 3 – 4 – 5 – 6 – 7 – Appealing
- Bold – 1 – 2 – 3 – 4 – 5 – 6 – 7 – Cautious
- Innovative – 1 – 2 – 3 – 4 – 5 – 6 – 7 – Conservative
- Dull – 1 – 2 – 3 – 4 – 5 – 6 – 7 – Captivating

T. Kojić, *User Experience for Serious Games in Virtual Reality*, T-Labs Series in Telecommunication Services, https://doi.org/10.1007/978-3-031-75530-9

- Undemanding – 1 – 2 – 3 – 4 – 5 – 6 – 7 – Challenging
- Motivating – 1 – 2 – 3 – 4 – 5 – 6 – 7 – Discouraging
- Novel – 1 – 2 – 3 – 4 – 5 – 6 – 7 – Ordinary
- Unruly – 1 – 2 – 3 – 4 – 5 – 6 – 7 – Manageable

SAM

Pleasure

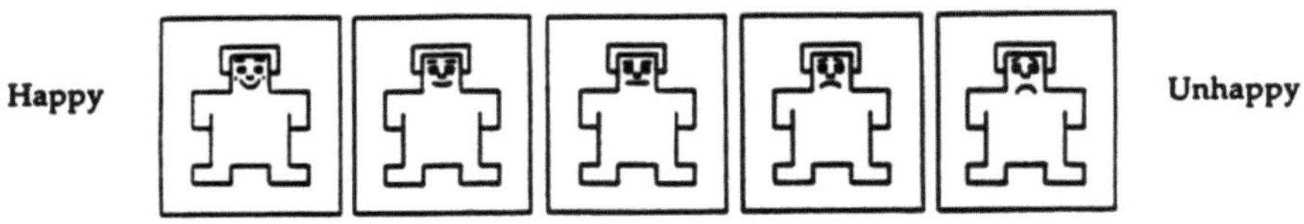

Arousal

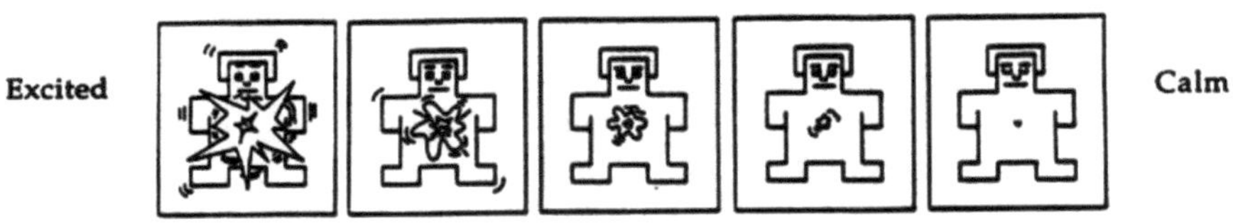

Dominance

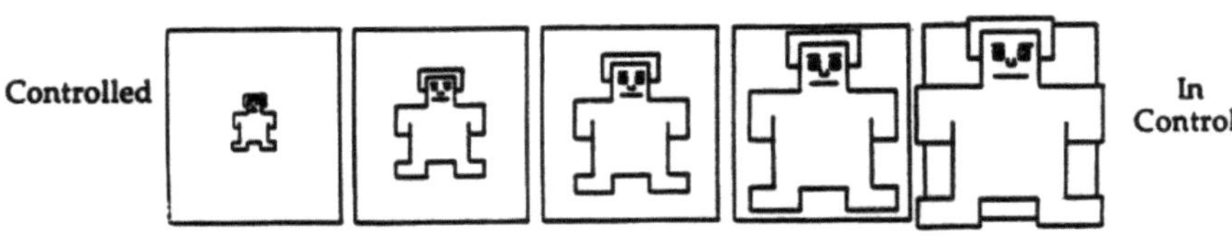

ATI

1. I like to occupy myself in greater detail with technical systems.
   - Completely disagree – 1 – 2 – 3 – 4 – 5 – 6 – Completely agree
2. I like testing the functions of new technical systems.
   - Completely disagree – 1 – 2 – 3 – 4 – 5 – 6 – Completely agree
3. I predominantly deal with technical systems because I have to.
   - Completely disagree – 1 – 2 – 3 – 4 – 5 – 6 – Completely agree

4. When I have a new technical system in front of me, I try it out intensively.
   - Completely disagree – 1 – 2 – 3 – 4 – 5 – 6 – Completely agree
5. I enjoy spending time becoming acquainted with a new technical system.
   - Completely disagree – 1 – 2 – 3 – 4 – 5 – 6 – Completely agree
6. It is enough for me that a technical system works; I do not care how or why.
   - Completely disagree – 1 – 2 – 3 – 4 – 5 – 6 – Completely agree
7. I try to understand how a technical system exactly works.
   - Completely disagree – 1 – 2 – 3 – 4 – 5 – 6 – Completely agree
8. It is enough for me to know the basic functions of a technical system.
   - Completely disagree – 1 – 2 – 3 – 4 – 5 – 6 – Completely agree
9. I try to make full use of the capabilities of a technical system.
   - Completely disagree – 1 – 2 – 3 – 4 – 5 – 6 – Completely agree

NMQ

1. I often felt as if my opponent and I were in the same virtual environment together.
   - Strongly disagree – 1 – 2 – 3 – 4 – 5 – Strongly agree
2. I think my opponent often felt as if we were in the same virtual environment together.
   - Strongly disagree – 1 – 2 – 3 – 4 – 5 – Strongly agree
3. I was often aware of my opponent in the virtual environment.
   - Strongly disagree – 1 – 2 – 3 – 4 – 5 – Strongly agree
4. My opponent was often aware of me in the virtual environment.
   - Strongly disagree – 1 – 2 – 3 – 4 – 5 – Strongly agree
5. I hardly noticed my opponent in the virtual environment
   - Strongly disagree – 1 – 2 – 3 – 4 – 5 – Strongly agree
6. My opponent did not notice me in the virtual environment.
   - Strongly disagree – 1 – 2 – 3 – 4 – 5 – Strongly agree
7. I often felt as if we were in different places rather than together in the same virtual environment.
   - Strongly disagree – 1 – 2 – 3 – 4 – 5 – Strongly agree

8. I think my opponent often felt as if we were in different places rather than together in the same virtual environment.
   - Strongly disagree – 1 – 2 – 3 – 4 – 5 – Strongly agree
9. I paid close attention to my opponent.
   - Strongly disagree – 1 – 2 – 3 – 4 – 5 – Strongly agree
10. My opponent paid close attention to me.
    - Strongly disagree – 1 – 2 – 3 – 4 – 5 – Strongly agree
11. I was easily distracted from my opponent when other things were going on.
    - Strongly disagree – 1 – 2 – 3 – 4 – 5 – Strongly agree
12. My opponent was easily distracted from me when other things were going on.
    - Strongly disagree – 1 – 2 – 3 – 4 – 5 – Strongly agree
13. I tended to ignore my opponent.
    - Strongly disagree – 1 – 2 – 3 – 4 – 5 – Strongly agree
14. My opponent tended to ignore me.
    - Strongly disagree – 1 – 2 – 3 – 4 – 5 – Strongly agree
15. I was sometimes influenced by my opponent's moods.
    - Strongly disagree – 1 – 2 – 3 – 4 – 5 – Strongly agree
16. My opponent was sometimes influenced by my moods.
    - Strongly disagree – 1 – 2 – 3 – 4 – 5 – Strongly agree
17. When I was happy, my opponent tended to be happy.
    - Strongly disagree – 1 – 2 – 3 – 4 – 5 – Strongly agree
18. When my opponent was happy, I tended to be happy.
    - Strongly disagree – 1 – 2 – 3 – 4 – 5 – Strongly agree
19. When I was feeling sad, my opponent also seemed to be down.
    - Strongly disagree – 1 – 2 – 3 – 4 – 5 – Strongly agree
20. When my opponent was feeling sad, I tended to be sad.
    - Strongly disagree – 1 – 2 – 3 – 4 – 5 – Strongly agree
21. When I was feeling nervous, my opponent also seemed to be nervous.
    - Strongly disagree – 1 – 2 – 3 – 4 – 5 – Strongly agree
22. When my opponent was nervous, my opponent I tended to be nervous.
    - Strongly disagree – 1 – 2 – 3 – 4 – 5 – Strongly agree

23. I was able to communicate my intentions clearly to my opponent.
    - Strongly disagree – 1 – 2 – 3 – 4 – 5 – Strongly agree
24. My opponent was able to communicate their intentions clearly to me.
    - Strongly disagree – 1 – 2 – 3 – 4 – 5 – Strongly agree
25. My thoughts were clear to my opponent.
    - Strongly disagree – 1 – 2 – 3 – 4 – 5 – Strongly agree
26. My opponent's thoughts were clear to me.
    - Strongly disagree – 1 – 2 – 3 – 4 – 5 – Strongly agree
27. I was able to understand what my opponent meant.
    - Strongly disagree – 1 – 2 – 3 – 4 – 5 – Strongly agree
28. My opponent was able to understand what I meant.
    - Strongly disagree – 1 – 2 – 3 – 4 – 5 – Strongly agree
29. My actions were often dependent on my opponent's actions.
    - Strongly disagree – 1 – 2 – 3 – 4 – 5 – Strongly agree
30. My opponent's actions were often dependent on my actions.
    - Strongly disagree – 1 – 2 – 3 – 4 – 5 – Strongly agree
31. My behavior was often in direct response to my opponent's behavior.
    - Strongly disagree – 1 – 2 – 3 – 4 – 5 – Strongly agree
32. The behavior of my opponent was often in direct response to my behavior.
    - Strongly disagree – 1 – 2 – 3 – 4 – 5 – Strongly agree
33. What I did often affected what my opponent did.
    - Strongly disagree – 1 – 2 – 3 – 4 – 5 – Strongly agree
34. What my opponent did often affected what I did.
    - Strongly disagree – 1 – 2 – 3 – 4 – 5 – Strongly agree

MPAM-R

1. Because I want to be physically fit.
    - Not at all true for me – 1 – 2 – 3 – 4 – 5 – 6 – 7 – Very true for me
2. Because it is fun.
    - Not at all true for me – 1 – 2 – 3 – 4 – 5 – 6 – 7 – Very true for me

3. Because I like engaging in activities which physically challenge me.
   - Not at all true for me – 1 – 2 – 3 – 4 – 5 – 6 – 7 – Very true for me
4. Because I want to obtain new skills.
   - Not at all true for me – 1 – 2 – 3 – 4 – 5 – 6 – 7 – Very true for me
5. Because I want to look or maintain weight, so I look better.
   - Not at all true for me – 1 – 2 – 3 – 4 – 5 – 6 – 7 – Very true for me
6. Because I want to be with my friends.
   - Not at all true for me – 1 – 2 – 3 – 4 – 5 – 6 – 7 – Very true for me
7. Because I like to do this activity.
   - Not at all true for me – 1 – 2 – 3 – 4 – 5 – 6 – 7 – Very true for me
8. Because I want to improve existing skills.
   - Not at all true for me – 1 – 2 – 3 – 4 – 5 – 6 – 7 – Very true for me
9. Because I like the challenge.
   - Not at all true for me – 1 – 2 – 3 – 4 – 5 – 6 – 7 – Very true for me
10. Because I want to define my muscles, so I look better.
    - Not at all true for me – 1 – 2 – 3 – 4 – 5 – 6 – 7 – Very true for me
11. Because it makes me happy.
    - Not at all true for me – 1 – 2 – 3 – 4 – 5 – 6 – 7 – Very true for me
12. Because I want to keep up my current skill level.
    - Not at all true for me – 1 – 2 – 3 – 4 – 5 – 6 – 7 – Very true for me
13. Because I want to have more energy.
    - Not at all true for me – 1 – 2 – 3 – 4 – 5 – 6 – 7 – Very true for me
14. Because I like activities that are physically challenging.
    - Not at all true for me – 1 – 2 – 3 – 4 – 5 – 6 – 7 – Very true for me
15. Because I like to be with others who are interested in this activity.
    - Not at all true for me – 1 – 2 – 3 – 4 – 5 – 6 – 7 – Very true for me
16. Because I want to improve my cardiovascular fitness.
    - Not at all true for me – 1 – 2 – 3 – 4 – 5 – 6 – 7 – Very true for me
17. Because I want to improve my appearance.
    - Not at all true for me – 1 – 2 – 3 – 4 – 5 – 6 – 7 – Very true for me

18. Because I think it is interesting.
    - Not at all true for me – 1 – 2 – 3 – 4 – 5 – 6 – 7 – Very true for me
19. Because I want to maintain my physical strength to live a healthy life.
    - Not at all true for me – 1 – 2 – 3 – 4 – 5 – 6 – 7 – Very true for me
20. Because I want to be attractive to others.
    - Not at all true for me – 1 – 2 – 3 – 4 – 5 – 6 – 7 – Very true for me
21. Because I want to meet new people.
    - Not at all true for me – 1 – 2 – 3 – 4 – 5 – 6 – 7 – Very true for me
22. Because I enjoy this activity.
    - Not at all true for me – 1 – 2 – 3 – 4 – 5 – 6 – 7 – Very true for me
23. Because I want to maintain my physical health and well-being.
    - Not at all true for me – 1 – 2 – 3 – 4 – 5 – 6 – 7 – Very true for me
24. Because I want to improve my body shape.
    - Not at all true for me – 1 – 2 – 3 – 4 – 5 – 6 – 7 – Very true for me
25. Because I want to get better at my activity.
    - Not at all true for me – 1 – 2 – 3 – 4 – 5 – 6 – 7 – Very true for me
26. Because I find this activity stimulating.
    - Not at all true for me – 1 – 2 – 3 – 4 – 5 – 6 – 7 – Very true for me
27. Because I will feel physically unattractive if I do not.
    - Not at all true for me – 1 – 2 – 3 – 4 – 5 – 6 – 7 – Very true for me
28. Because my friends want me to.
    - Not at all true for me – 1 – 2 – 3 – 4 – 5 – 6 – 7 – Very true for me
29. Because I like the excitement of participation.
    - Not at all true for me – 1 – 2 – 3 – 4 – 5 – 6 – 7 – Very true for me
30. Because I enjoy spending time with others doing this activity.
    - Not at all true for me – 1 – 2 – 3 – 4 – 5 – 6 – 7 – Very true for me

USE

1. It helps me be more effective.
    - Strongly disagree – 1 – 2 – 3 – 4 – 5 – 6 – 7 – Strongly agree

2. It helps me be more productive.
   - Strongly disagree – 1 – 2 – 3 – 4 – 5 – 6 – 7 – Strongly agree
3. It is useful.
   - Strongly disagree – 1 – 2 – 3 – 4 – 5 – 6 – 7 – Strongly agree
4. It gives me more control over the activities in my life.
   - Strongly disagree – 1 – 2 – 3 – 4 – 5 – 6 – 7 – Strongly agree
5. It makes the things I want to accomplish easier to get done.
   - Strongly disagree – 1 – 2 – 3 – 4 – 5 – 6 – 7 – Strongly agree
6. It saves me time when I use it.
   - Strongly disagree – 1 – 2 – 3 – 4 – 5 – 6 – 7 – Strongly agree
7. It meets my needs.
   - Strongly disagree – 1 – 2 – 3 – 4 – 5 – 6 – 7 – Strongly agree
8. It does everything I would expect it to do.
   - Strongly disagree – 1 – 2 – 3 – 4 – 5 – 6 – 7 – Strongly agree
9. It is easy to use.
   - Strongly disagree – 1 – 2 – 3 – 4 – 5 – 6 – 7 – Strongly agree
10. It is simple to use.
    - Strongly disagree – 1 – 2 – 3 – 4 – 5 – 6 – 7 – Strongly agree
11. It is user friendly.
    - Strongly disagree – 1 – 2 – 3 – 4 – 5 – 6 – 7 – Strongly agree
12. It requires the fewest steps possible to accomplish what I want to do with it.
    - Strongly disagree – 1 – 2 – 3 – 4 – 5 – 6 – 7 – Strongly agree
13. It is flexible.
    - Strongly disagree – 1 – 2 – 3 – 4 – 5 – 6 – 7 – Strongly agree
14. Using it is effortless.
    - Strongly disagree – 1 – 2 – 3 – 4 – 5 – 6 – 7 – Strongly agree
15. I can use it without written instructions.
    - Strongly disagree – 1 – 2 – 3 – 4 – 5 – 6 – 7 – Strongly agree
16. I do not notice any inconsistencies as I use it.
    - Strongly disagree – 1 – 2 – 3 – 4 – 5 – 6 – 7 – Strongly agree

17. Both occasional and regular users would like it.
    - Strongly disagree – 1 – 2 – 3 – 4 – 5 – 6 – 7 – Strongly agree
18. I can recover from mistakes quickly and easily.
    - Strongly disagree – 1 – 2 – 3 – 4 – 5 – 6 – 7 – Strongly agree
19. I can use it successfully every time.
    - Strongly disagree – 1 – 2 – 3 – 4 – 5 – 6 – 7 – Strongly agree
20. I learned to use it quickly.
    - Strongly disagree – 1 – 2 – 3 – 4 – 5 – 6 – 7 – Strongly agree
21. I easily remember how to use it.
    - Strongly disagree – 1 – 2 – 3 – 4 – 5 – 6 – 7 – Strongly agree
22. It is easy to learn to use it.
    - Strongly disagree – 1 – 2 – 3 – 4 – 5 – 6 – 7 – Strongly agree
23. I quickly became skillful with it.
    - Strongly disagree – 1 – 2 – 3 – 4 – 5 – 6 – 7 – Strongly agree
24. I am satisfied with it.
    - Strongly disagree – 1 – 2 – 3 – 4 – 5 – 6 – 7 – Strongly agree
25. I would recommend it to a friend.
    - Strongly disagree – 1 – 2 – 3 – 4 – 5 – 6 – 7 – Strongly agree
26. It is fun to use.
    - Strongly disagree – 1 – 2 – 3 – 4 – 5 – 6 – 7 – Strongly agree
27. It works the way I want it to work.
    - Strongly disagree – 1 – 2 – 3 – 4 – 5 – 6 – 7 – Strongly agree
28. It is wonderful.
    - Strongly disagree – 1 – 2 – 3 – 4 – 5 – 6 – 7 – Strongly agree
29. I feel I need to have it.
    - Strongly disagree – 1 – 2 – 3 – 4 – 5 – 6 – 7 – Strongly agree
30. It is pleasant to use.
    - Strongly disagree – 1 – 2 – 3 – 4 – 5 – 6 – 7 – Strongly agree

UEQ-S

This document lists the items and their order for all the available languages of the UEQ.

- Obstructive – 1 – 2 – 3 – 4 – 5 – 6 – 7 – Supportive
- Complicated – 1 – 2 – 3 – 4 – 5 – 6 – 7 – Easy
- Inefficient – 1 – 2 – 3 – 4 – 5 – 6 – 7 – Efficient
- Confusing – 1 – 2 – 3 – 4 – 5 – 6 – 7 – Clear
- Boring – 1 – 2 – 3 – 4 – 5 – 6 – 7 – Exciting
- Not interesting – 1 – 2 – 3 – 4 – 5 – 6 – 7 – Interesting
- Conventional – 1 – 2 – 3 – 4 – 5 – 6 – 7 – Inventive
- Usual – 1 – 2 – 3 – 4 – 5 – 6 – 7 – Leading edge

SUS

1. I think that I would like to use hand tracking for VR interactions frequently.
   - Strongly disagree – 1 – 2 – 3 – 4 – 5 – Strongly agree
2. I found hand tracking for VR interactions unnecessarily complex.
   - Strongly disagree – 1 – 2 – 3 – 4 – 5 – Strongly agree
3. I thought hand tracking for VR interactions was easy to use.
   - Strongly disagree – 1 – 2 – 3 – 4 – 5 – Strongly agree
4. I think that I would need the support of a technical person to be able to use hand tracking for VR interactions.
   - Strongly disagree – 1 – 2 – 3 – 4 – 5 – Strongly agree
5. I found the various functions in hand tracking for VR interactions well integrated.
   - Strongly disagree – 1 – 2 – 3 – 4 – 5 – Strongly agree
6. I thought there was too much inconsistency in hand tracking for VR interactions.
   - Strongly disagree – 1 – 2 – 3 – 4 – 5 – Strongly agree
7. I would imagine that most people would learn to use hand tracking for VR interactions very quickly.
   - Strongly disagree – 1 – 2 – 3 – 4 – 5 – Strongly agree
8. I found hand tracking for VR interactions very cumbersome to use.
   - Strongly disagree – 1 – 2 – 3 – 4 – 5 – Strongly agree

9. I felt very confident using hand tracking for VR interactions.
   - Strongly disagree – 1 – 2 – 3 – 4 – 5 – Strongly agree
10. I needed to learn a lot of things before I could get going with hand tracking for VR interactions.
    - Strongly disagree – 1 – 2 – 3 – 4 – 5 – Strongly agree

IPQ

1. In the computer generated world I had a sense of "being there"
   - Not at all– 1 – 2 – 3 – 4 – 5 – Very much
2. How real did the virtual world seem to you?
   - Not real at all– 1 – 2 – 3 – 4 – 5 – Completely real
3. How much did your experience in the virtual environment seem consistent with your real-world experience ?
   - Not consistent– 1 – 2 – 3 – 4 – 5 – Very consistent
4. How real did the virtual world seem to you?
   - About as real as an imagined world– 1 – 2 – 3 – 4 – 5 – Indistinguishable from the real world
5. The virtual world seemed more realistic than the real world.
   - Fully disagree– 1 – 2 – 3 – 4 – 5 – Fully agree

S FSS

1. I felt I was competent enough to meet the demands of the situation.
   - Strongly disagree – 1 – 2 – 3 – 4 – 5 – Strongly agree
2. I did things spontaneously and automatically without having to think.
   - Strongly disagree – 1 – 2 – 3 – 4 – 5 – Strongly agree
3. I had a strong sense of what I wanted to do.
   - Strongly disagree – 1 – 2 – 3 – 4 – 5 – Strongly agree
4. I had a good idea about how well I was doing while I was involved in the task-activity.
   - Strongly disagree – 1 – 2 – 3 – 4 – 5 – Strongly agree

5. I was completely focused on the task at hand.
   - Strongly disagree – 1 – 2 – 3 – 4 – 5 – Strongly agree
6. I had a feeling of total control over what I was doing.
   - Strongly disagree – 1 – 2 – 3 – 4 – 5 – Strongly agree
7. I was not worried about what others may have been thinking of me.
   - Strongly disagree – 1 – 2 – 3 – 4 – 5 – Strongly agree
8. The way time passed seemed to be different from normal.
   - Strongly disagree – 1 – 2 – 3 – 4 – 5 – Strongly agree
9. I found the experience extremely rewarding.
   - Strongly disagree – 1 – 2 – 3 – 4 – 5 – Strongly agree

SAQ

1. It felt appropriate to use the VR set in the shown public setting.
   - Not at all – 1 – 2 – 3 – 4 – 5 – 6 – 7 – Very much
2. It felt rude to use the VR set in the shown public setting.
   - Not at all – 1 – 2 – 3 – 4 – 5 – 6 – 7 – Very much
3. It felt uncomfortable being watched while using the VR in the shown public setting.
   - Not at all – 1 – 2 – 3 – 4 – 5 – 6 – 7 – Very much
4. I think the VR set makes me look cool.
   - Not at all – 1 – 2 – 3 – 4 – 5 – 6 – 7 – Very much
5. It would be useful for me if the people around me could communicate with me.
   - Not at all – 1 – 2 – 3 – 4 – 5 – 6 – 7 – Very much
6. It felt awkward doing head movements while using VR in public.
   - Not at all – 1 – 2 – 3 – 4 – 5 – 6 – 7 – Very much
7. It felt awkward performing body movements and hand gestures while using VR in public.
   - Not at all – 1 – 2 – 3 – 4 – 5 – 6 – 7 – Very much
8. I did not like the fact that I was isolated from the rest of the people.
   - Not at all – 1 – 2 – 3 – 4 – 5 – 6 – 7 – Very much

9. It would be interesting if the other people could see what I was doing and seeing in the VR.
   - Not at all – 1 – 2 – 3 – 4 – 5 – 6 – 7 – Very much
10. I was concerned about spectators recording me while using VR.
    - Not at all – 1 – 2 – 3 – 4 – 5 – 6 – 7 – Very much
11. I was concerned about bumping to objects and people while using the VR in the shown public setting.
    - Not at all – 1 – 2 – 3 – 4 – 5 – 6 – 7 – Very much

Appendix B
Additional Questionnaires

Demographics Questionnaire (US1 and US2)

1. What is your gender?
   - Female
   - Male
   - Other
2. What is your age?
3. What is your profession?
4. How often do you do sports?
   - 1x a month
   - 2x or more a month
   - 1x a week
   - 2–3x a week
   - 4–5x a week
   - Nearly every day
5. On average, how many hours a week do you personally spend on doing sport?
6. How experienced are you with Virtual Reality technologies (Cave, Head-Mounted Displays, e.g., HTC Vive)?
   - Not at all – 1 – 2 – 3 – 4 – 5 – Very experienced
7. How experienced are you in rowing?
   - Not at all – 1 – 2 – 3 – 4 – 5 – Very experienced

T. Kojić, *User Experience for Serious Games in Virtual Reality*, T-Labs Series in Telecommunication Services, https://doi.org/10.1007/978-3-031-75530-9

Post-test Questionnaire (US1)

1. Which session did you enjoy the most?
   - Without VR and without audio
   - Without VR but with audio
   - With VR but without audio
   - With VR and with audio
2. Why?
3. How would you evaluate your performance?
   - Very bad – 1 – 2 – 3 – 4 – 5 – Very good
4. Feel free to leave any other comments.

Post-condition Questionnaire (US2)

1. How readable were the metrics?
   - Not at all readable – 1 – 2 – 3 – 4 – 5 – 6 – 7 – Very readable
2. It was clear to me what the metrics represent.
   - Not at all true – 1 – 2 – 3 – 4 – 5 – 6 – 7 – Very true
3. The metrics have supported me in the workout.
   - Not at all true – 1 – 2 – 3 – 4 – 5 – 6 – 7 – Very true
4. I succeeded at maintaining the same speed throughout the whole workout.
   - Not at all true – 1 – 2 – 3 – 4 – 5 – 6 – 7 – Very true

Post-test Questionnaire (US2)

1. Which session did you enjoy the most?
   - Cockpit with gamified UI
   - Cockpit with descriptive UI (numbers)
   - Boat with gamified UI
   - Boat with descriptive UI (numbers)
2. Why?
3. How would you evaluate your performance?
   - Very bad – 1 – 2 – 3 – 4 – 5 – Very good
4. Would you need some other metrics to be shown during workout?
5. Feel free to leave any other comments.

Demographics Questionnaire (US3)

1. What is your gender?
   - Female
   - Male
   - Other
2. How old are you?
3. How often do you play multiplayer video games?
   - Never – 1 – 2 – 3 – 4 – 5 – Very often

Post-condition Questionnaire (US3)

1. Rate the lag of your boat.
   - No lag at all – 1 – 2 – 3 – 4 – 5 – 6 – 7 – 8 – 9 – 10 – Very strong lag
2. Rate the lag of your opponent's boat.
   - No lag at all – 1 – 2 – 3 – 4 – 5 – 6 – 7 – 8 – 9 – 10 – Very strong lag
3. Rate the overall lag in the virtual environment.
   - No lag at all – 1 – 2 – 3 – 4 – 5 – 6 – 7 – 8 – 9 – 10 – Very strong lag
4. Was the lag of your boat acceptable?
   - Very acceptable – 1 – 2 – 3 – 4 – 5 – 6 – 7 – 8 – 9 – 10 – Completely unacceptable

Demographics Questionnaire (US4)

1. What is your age?
2. What is your gender?
   - Female
   - Male
   - Other
3. How experienced are you in rowing and/or indoor rowing?
   - No experience at all – 1 – 2 – 3 – 4 – 5 – 6 – 7 – Very experienced
4. How many hours did you sleep last night?
5. Please rate your current mood.

Post-condition Questionnaire (US4)

1. I think I solved the synchronization task very well.
   - Strongly Disagree – 1 – 2 – 3 – 4 – 5 – 6 – 7 – Strongly Agree
2. It was hard for me to synchronize the breathing pattern with my rowing strokes.
   - Strongly Disagree – 1 – 2 – 3 – 4 – 5 – 6 – 7 – Strongly Agree
3. The chart helped me to sustain my breathing rhythm during rowing.
   - Strongly Disagree – 1 – 2 – 3 – 4 – 5 – 6 – 7 – Strongly Agree
4. The lung animation helped me to sustain my breathing rhythm during rowing.
   - Strongly Disagree – 1 – 2 – 3 – 4 – 5 – 6 – 7 – Strongly Agree
5. The synchronization rate in percent helped me to sustain my breathing rhythm during rowing.
   - Strongly Disagree – 1 – 2 – 3 – 4 – 5 – 6 – 7 – Strongly Agree
6. The textual synchronization feedback helped me to sustain my breathing rhythm during rowing.
   - Strongly Disagree – 1 – 2 – 3 – 4 – 5 – 6 – 7 – Strongly Agree

Post-test Questionnaire (US4)

1. Which interface did you like the most? Please rank the following interfaces from the highest (1) to the lowest (5):
   - The chart
   - The lung animation
   - The synchronization rate in percent
   - The textual synchronization feedback
   - Without VR interface
2. Which interface was the most helpful one for keeping the recommended breathing rhythm? Please rank the following interfaces from the highest (1) to the lowest (5):
   - The chart
   - The lung animation

- The synchronization rate in percent
- The textual synchronization feedback
- Without VR interface

Demographics Questionnaire (US5)

1. What is your gender?
   - Female
   - Male
   - Other
2. What is your age?
3. What is your profession?
4. How experienced are you with Virtual Reality technologies (Cave, Head-Mounted Displays, e.g., HTC Vive)?
   - Not at all – 1 – 2 – 3 – 4 – 5 – Very experienced
5. On average, how many hours do you spend reading (e.g., books, e-books, news, blog articles) per day?
   - Less than or equal to 1 hour
   - Between 2 and 4 hours
   - Between 5 and 7 hours
   - Between 8 and 10 hours
   - More than or equal to 11 hours
6. Do you use any form of eyesight improvement?
   - None
   - Glasses
   - Lenses
   - Other

Post-test Questionnaire (US5)

1. Which parameter was the easier for you to set up?
   - Size of text
   - Contrast of text
   - Distance of text
2. Why?
3. How would you evaluate readability of texts in VR?
   - Very bad – 1 – 2 – 3 – 4 – 5 – Very good

4. Can you imagine reading text in future inside of VR?
   - Yes
   - No
5. Did you feel like something was missing in the VR sessions?
   - Yes
   - No
6. If yes, what?
7. Do you have suggestions or comments about readability of text in VR?
8. Feel free to leave any other comments.

Demographics Questionnaire (US6)

1. What is your gender?
   - Female
   - Male
   - Other
2. What is your age?
3. What is your profession?
   - Student
   - Employed
   - Not working or studying at the moment
   - Retired
4. How experienced are you with Virtual Reality technologies (Cave, Head-Mounted Displays, e.g., HTC Vive)?
   - Not at all – 1 – 2 – 3 – 4 – 5 – Very experienced
5. How experienced are you with using hand tracking technologies?
   - Not at all – 1 – 2 – 3 – 4 – 5 – Very experienced

Post-test Questionnaire (US6)

1. Please rate your preferred way of interacting as worst, bad, good, and best.
   - Hand tracking
   - Controllers and hands
   - Controllers and no hands
   - Controllers only hands

2. How would you evaluate hand tracking for VR interactions?
   - Very bad – 1 – 2 – 3 – 4 – 5 – Very good
3. Can you imagine using hand tracking for VR interactions in the future?
   - Yes
   - No
4. Do you have any additional comments?
   - Yes
   - No
5. If yes, what?

Demographics Questionnaire (US7)

1. What is your gender?
   - Female
   - Male
   - Other
   - Prefer not to answer
2. Please select the category that includes your age (years).
   - Under 18
   - 18–24
   - 25–34
   - 35–44
   - 45–54
   - 55 or above
   - Prefer not to answer
3. What best describes your marital status?
   - Single
   - Married
   - Prefer not to answer
4. What best describes your employment status?
   - Student
   - Working student
   - Part-time employment
   - Full-time employment
   - Not employed, but looking for work
   - Not employed and not looking for work

- Self-employed
- Retired
- Prefer not to answer

5. Profession (also answer if working + studying)
6. What best describes your level of education?
   - High school graduate or equivalent
   - High school (in progress)
   - Bachelor's degree
   - Bachelor's degree (in progress)
   - Masters' degree
   - Masters' degree (in progress)
   - Postgraduate or higher
   - Postgraduate or higher degree (in progress)
   - None of the above
7. Which country and city do you currently live in (i.e., Berlin, Germany)?
8. How often do you withdraw money from an ATM (per month)?
   - Never
   - 1–3 times
   - 4–5 times
   - 6–10 times
   - More than 10 times
9. How often do you check bank/financial statement status (per month)?
   - Never
   - 1–3 times
   - 4–5 times
   - 6–10 times
   - More than 10 times
10. How experienced are you in dealing with your own finances?
    - Not at all – 1 – 2 – 3 – 4 – 5 – Very experienced
11. Are you experienced with Virtual Reality technologies (Cave, Head-Mounted Displays, e.g., HTC Vive)?
    - Not at all – 1 – 2 – 3 – 4 – 5 – Very experienced
12. Have you been immersed in Virtual Reality before?
    - Yes
    - No

Demographics Questionnaire (US8)

1. Gender
   - Male – 1 – 2 – 3 – 4 – 5 – 6 – Female
2. Age
3. Nationality
4. Course of study/profession
5. What is your dominant hand?
   - Right
   - Left
   - Both
6. How tall are you? (in cm)
7. Have you used a VR headset before?
   - Yes
   - No
8. How familiar are you with VR technologies?
   - Not familiar at all – 1 – 2 – 3 – 4 – 5 – I know a lot

Post-test Questionnaire (US8)

1. What do you think of the graphics of the experience?
   - Bad – 1 – 2 – 3 – 4 – 5 – 6 – 7 – Good
2. Do you think 360 videos are a good tool to simulate real-life situation?
   - Yes
   - No
3. Which activity was the best for you?
   - Static game
   - Dynamic game
4. Why?
5. Which social environment have you liked the most?
   - One distant person
   - Few distant persons
   - Few close persons
   - Many close persons
6. Why?

7. In which social environment it was most appropriate to use VR?
    - One distant person
    - Few distant persons
    - Few close persons
    - Many close persons
8. Why?
9. Did the presence of others in the social environments affect your experience?
10. What is your opinion on moving your head, body, and hands in public while using VR?
11. Would you use VR in a public space?
    - Yes
    - No
    - Maybe
12. Why?
13. Which kind of experiences in VR would you feel comfortable to play in public places?
14. How do you feel about the overall experience of the experiment?
15. Is there anything else you want to share?

Appendix C
Additional Overview for Methodology or Results

Table C.1 Mean values of angular size and contrast ration over all conditions for the best settings of text parameters. Difference in conditions was by changing length of text—2 words (short), 21 words (medium), 51 words (long), and device (Oculus Go and Quest)

		Angular size		Contrast ratio	
Text	Device	Mean	SD	Mean	SD
Short	Go	27.48	9.86	11.89	6.68
Short	Quest	32.06	18.75	10.12	7.32
Medium	Go	17.17	4.90	10.47	7.52
Medium	Quest	16.18	5.25	8.71	5.94
Long	Go	16.01	9.25	11.02	7.73
Long	Quest	17.87	24.52	9.37	5.75

T. Kojić, *User Experience for Serious Games in Virtual Reality*, T-Labs Series in Telecommunication Services, https://doi.org/10.1007/978-3-031-75530-9

Table C.2 A summary of the methodology for all user studies as all in one overview

	Study 1	Study 2	Study 3	Study 4
Set-up				
Hardware	HTC Vive with PC, a wireless in-ear headset, a large TV and a Augletics Eight ergometer	HTC Vive with PC and a Augletics Eight ergometer	HTC Vive with PC and a Concept2 Model D rowing ergometer	HTC Vive with PC, a Augletics Eight ergometer and a BreathZpot sensor
Software	Rowing VR App	Rowing VR App	Rowing VR App	Rowing VR App
Conditions	VR and DTC: with the VE in VR and conversation possibility, VR without DTC: with the VE in VR but without the possibility of conversation with another participant, No VR and with DTC: with the TV and the possibility of conversation with another participant, No VR and no DTC: with the TV and without the possibility of conversation with another participant.	Gamified cockpit: at the front of the rowing scull with gamified visualization of metrics, Gamified coach boat: that follows the player with gamified visualization of metrics on a screen in the VR, Digitalized cockpit: at the front of the rowing scull with digital visualization of metrics, Digitalized coach boat: that follows the player with digital visualization of metrics on a screen in the VR.	Own L and Opp. 's L: as player's low (30ms) and opponent's low (30ms) Own L and Opp. 's M: as player's low (30ms) and opponent's medium (100ms) Own L and Opp. 's H: as player's low (30ms) and opponent's high (500ms) Own M and Opp. 's L: as player's medium (100ms) and opponent's low (30ms) Own M and Opp. 's M: as player's medium (100ms) and opponent's medium (100ms) Own H and Opp. 's L: as player's high (500ms) and opponent's low (30ms) Own H and Opp. 's H: as player's high (500ms) and opponent's high (500ms)	No VR: an option that is not using VR and without breathing biofeedback, VR Only: an option in VR but without breathing biofeedback, VR Chart: constant breathing biofeedback as line chart and synchronization display in VR, VR Lung: constant breathing biofeedback as lung animation and synchronization display in VR, VR All: all visualizations of constant breathing biofeedback displayed in VR.
Task	Rowing against the other player in live race.	Task to choose a comfortable speed in kilometers per hour, which they would be able to maintain throughout all four workouts.	Rowing agains the artifical opponent with different delay settings	Maintain correct breathing rythem to breath out when the pull starts.
Parameters	Demographics Flow (S FSS-2) Presence (IPQ) Social presence (NMQ) Preferred setup	Demographics General motivation for sports (Mpam-r) Flow (S FSS-2) User Experience (UEQ-S) Readability and clearness of UI (7-point Lickert scale)	Demographics QoE for the delay of oneself and of the opponent (10-point Likert scale) Presence (IPQ) Flow (S FSS-2)	Demographics Flow (S FSS-2) Presence (IPQ) User Experience (AttrakDiff) Ranking different vizualizations by sympathy and helpfullness
Participants				
Number	36 participants in lab study	26 participants in lab study	23 participants in lab study	23 participants in lab study
Demographics	20 female and 16 male. The average age of the participants was 28.16 years (SD = 6.14, min = 18, max = 47).	16 male and 10 female. The average age of the participants was 28.54 years (SD = 7.38 , min = 18, max = 54).	17 male and 6 female. The average recorded age was 25.57 years (SD = 3.41 years).	10 female and 13 male. The mean age was 23.67 years (SD = 4.887, min age = 17, max = 36).

	Study 5	Study 6	Study 7	Study 8
Set-up				
Hardware	Oculus Go, Oculus Quest	Oculus Quest	Oculus Go	Oculus Quest
Software	Readability VR App	Haptics App	Finance App	Simulated Social VR app
Conditions	Go and Short: showing the short text of 2 words, Go and Medium: showing the medium text of 21 words, Go and Long: showing the long text of 51 words, Quest and Short: showing short text of 2 words, Quest and Medium: showing medium text of 21 words, Quest and Long: showing long text of 51 words.	C_H: visualized controllers and hands, C_noH: visualized controllers, but no visualization of hands, C_OnlyH: visualized hands, but no visualization of controllers, HTrack: visualized hands while using hand tracking.	Personal finance statement as 3D vizualization in VR Personal finance statement as 2D vizualization in paperwork	One Distant Static: one distant person and static game Few Distant Static: few distant persons and static game Few Close Static: few close persons and static game Many Close Static: many close persons and static game One Distant Dynamic: distant person and dynamic game Few Distant Dynamic: few distant persons, dynamic game Few Close Dynamic: few close persons and dynamic game Many Close Dynamic: many close persons, dynamic game
Task	For each condition: set text parameters to get the best and the worst possible readability	Sorting objects by color and retyping number sequence.	Anwser questions (such as "Find the quarter in which the least amount of expenses was spent.") about the finance situation based on given data.	Select numbers in the correct order from 1 to 50 for static game, and grabbing, dodging and shielding from objects in dynamic game.
Parameters	Angular size (dmm) Contrast ratio Afinity for Technology (ATI) Emotional assesment (SAM) System Usability (SUS)	Demographics Affinity for technology (ATI) Emotional assesment (SAM) Presence (IPQ) System Usability (SUS)	Demographics Presence (IPQ) User Satisfaction (USE) System Usability (SUS)	Demographics Affinity for technology (ATI) Presence (IPQ) User Experience (UEQ-S) Emotional assesment (SAM) Social acceptability (SAQ) System Usability (SUS)
Participants				
Number	22 participants in lab study	20 participants in lab study	23 participants in lab study	28 participats in lab study
Demographics	54.5% male and 45.5% female. The average age of the participants was 28.41 years (SD = 9.56 , min = 19, max = 62).	8 female and 12 male, with the average age of the participants was 30.20 years (SD = 12.74, min = 19, max = 62).	5 females and 18 males within different age group of range from 18 to 54 years old.	21 male and 7 female. The average age was 24.64 years (SD = 2.6, min = 21, max = 31).

References

1. Bhowmik, A.K.: Recent developments in virtual-reality and augmented-reality technologies. Inf. Display **33**(6), 20–32 (2017)
2. Berg, L.P., Vance, J.M.: Industry use of virtual reality in product design and manufacturing: a survey. Virtual Real. **21**(1), 1–17 (2017)
3. Velev, D., Zlateva, P.: Virtual reality challenges in education and training. Int. J. Learn. Teach. **3**(1), 33–37 (2017)
4. Alsop, T.: Consumer and Enterprise VR Revenue Worldwide 2024 (2021). [Online] Available: https://www.statista.com/statistics/1221522/virtual-reality-market-size-worldwide/
5. Bastug, E., Bennis, M., Médard, M., Debbah, M.: Toward interconnected virtual reality: opportunities, challenges, and enablers. IEEE Commun. Mag. **55**(6), 110–117 (2017)
6. Werbach, K., Hunter, D.: For the Win: How Game Thinking Can Revolutionize Your Business. Wharton Digital Press, Philadelphia (2012)
7. Deterding, S., Dixon, D., Khaled, R., Nacke, L.: From game design elements to gamefulness: defining gamification. In: Proceedings of the 15th International Academic MindTrek Conference: Envisioning Future Media Environments, pp. 9–15. ACM, New York (2011)
8. Bitkom: Durchschnittliche Spieldauer (Computer- und Videospiele) von Kindern und Jugendlichen in Deutschland im Jahr 2017. In: Statista—Das Statistik-Portal (2017). [Online]. Available: https://de.statista.com/statistik/daten/studie/306901/umfrage/spieldauer-von-kindern-und-jugendlichen-in-deutschland-nach-alter/ (Visited on 01/18/2018)
9. Hettinger, L.J., Riccio, G.E.: Visually induced motion sickness in virtual environments. Presence: Teleoper. Virtual Environ. **1**(3), 306–310 (1992)
10. Somrak, A., Humar, I., Hossain, M.S., Alhamid, M.F., Hossain, M.A., Guna, J.: Estimating VR sickness and user experience using different HMD technologies: an evaluation study. Future Gener. Comput. Syst. **94**, 302–316 (2019)
11. Lin, J.-W., Duh, H.B.-L., Parker, D.E., Abi-Rached, H., Furness, T.A.: Effects of field of view on presence, enjoyment, memory, and simulator sickness in a virtual environment. In: Proceedings IEEE Virtual Reality 2002, pp. 164–171. IEEE, Piscatway (2002)
12. Ashtari, N., Bunt, A., McGrenere, J., Nebeling, M., Chilana, P. K.: Creating augmented and virtual reality applications: Current practices, challenges, and opportunities. In: Proceedings of the 2020 CHI Conference on Human Factors in Computing Systems, pp. 1–13 (2020)
13. Garrett, J.J.: The Elements of User Experience: User-Centered Design for the Web and Beyond. Pearson Education, London (2010)
14. Kuniavsky, M.: Observing the User Experience: A Practitioner's Guide to User Research. Elsevier, Amsterdam (2003)

T. Kojić, *User Experience for Serious Games in Virtual Reality*, T-Labs Series in Telecommunication Services, https://doi.org/10.1007/978-3-031-75530-9

15. ITU-T Recommendation G.1035: Influencing Factors on Quality of Experience for Virtual Reality Services. International Telecommunication Union, Geneva (2021)
16. QUALINET White Paper on Definitions of Immersive Media Experience (IMEx): European Network on Quality of Experience in Multimedia Systems and Services, 14th QUALINET meeting. Perkis, A., Timmerer, C. (Eds.) (2020). arxiv.org/abs/2007.07032
17. Schmidt, S., Ehrenbrink, P., Weiss, B., Voigt-Antons, J.-N., Kojic, T., Johnston, A., Möller, S.: Impact of virtual environments on motivation and engagement during exergames. In: 2018 Tenth International Conference on Quality of Multimedia Experience (QoMEX), pp. 1–6. IEEE, Piscatway (2018)
18. Kojić, T., Voigt-Antons, J.-N., Schmidt, S., Tetzlaff, L., Kortowski, B., Sirotina, U., Möller, S.: Influence of virtual environments and conversations on user engagement during multiplayer exergames. In: 2018 Tenth International Conference on Quality of Multimedia Experience (QoMEX), pp. 1–3. IEEE, Piscatway (2018)
19. Kojic, T., Sirotina, U., Möller, S., Voigt-Antons, J.-N.: Influence of UI complexity and positioning on user experience during VR exergames. In: 2019 Eleventh International Conference on Quality of Multimedia Experience (QoMEX), pp. 1–6. IEEE, Piscatway (2019)
20. Kojic, T., Schmidt, S., Möller, S., Voigt-Antons, J.-N.: Influence of network delay in virtual reality multiplayer exergames: Who is actually delayed? In: 2019 Eleventh International Conference on Quality of Multimedia Experience (QoMEX), pp. 1–3. IEEE, Piscatway (2019)
21. Kojić, T., Nugyen, L.T., Voigt-Antons, J.-N.: Impact of constant visual biofeedback on user experience in virtual reality exergames. In: IEEE International Symposium on Multimedia (ISM), vol. 2019, pp. 307–3073. IEEE, Piscatway (2019)
22. Kojić, T., Ali, D., Greinacher, R., Möller, S., Voigt-Antons, J.-N.: User experience of reading in virtual reality—finding values for text distance, size and contrast. In: 2020 Twelfth International Conference on Quality of Multimedia Experience (QoMEX), pp. 1–6. IEEE, Piscatway (2020)
23. Voigt-Antons, J.-N., Kojic, T., Ali, D., Möller, S.: Influence of hand tracking as a way of interaction in virtual reality on user experience. In: 2020 Twelfth International Conference on Quality of Multimedia Experience (QoMEX), pp. 1–4. IEEE, Piscatway (2020)
24. Kojić, T., Ashipala, S., Möller, S., Voigt-Antons, J.-N.: Exploring visualisations for financial statements in virtual reality. In: 2020 IEEE International Conference on Artificial Intelligence and Virtual Reality (AIVR), pp. 49–52. IEEE, Piscatway (2020)
25. Vergari, M., Kojić, T., Vona, F., Garzotto, F., Möller, S., Voigt-Antons, J.-N.: Influence of interactivity and social environments on user experience and social acceptability in virtual reality. In: IEEE Virtual Reality and 3D User Interfaces (VR), vol. 2021, pp. 695–704. IEEE, Piscatway (2021)
26. T. Kojić, Voigt-Antons, J.-N., Möller, S.: Contribution to ITU-T Rec G.1035: Contribution for Human Influencing Factors—Static and Dynamic Human Factors," ITU-T Study Group 12, Virtual-Geneva, ITU-T Contribution C.562 (2021)
27. Kojić, T., Voigt-Antons, J.-N., Möller, S.: Update of Human IF for ITU-T Rec G.1035. ITU-T Study Group 12, Virtual-Geneva, ITU-T Contribution C.605 (2021)
28. Kojić, T., Voigt-Antons, J.-N., Möller, S.: Proposal for new Work Item P.IntVR: Subjective Test Methods for Interactive Virtual Reality Applications. ITU-T Study Group 12, Virtual-Geneva, ITU-T Contribution C.548 (2021)
29. Kojić, T., Voigt-Antons, J.-N., Möller, S.: P.IntVR: Subjective Test Method for Interactive Virtual Reality Applications," ITU-T Study Group 12, Virtual-Geneva, ITU-T Contribution C.601 (2021)
30. Kojić, T., Voigt-Antons, J.-N., Möller, S.: ITU-T P.IntVR: Subjective Test Method for Interactive Virtual Reality Applications–Quality of Experience aspects for Interactive Virtual Reality, ITU-T Study Group 12, Geneva, ITU-T Contribution C.060 (2022)

31. Sutherland, I.: The Ultimate Display (1965)
32. Lanier, J.: Dawn of the New Everything: Encounters with Reality and Virtual Reality. Henry Holt and Company, New York (2017)
33. Cummings, J.J., Bailenson, J.N.: How immersive is enough? A meta-analysis of the effect of immersive technology on user presence. Media Psychol. **19**(2), 272–309 (2016)
34. Steuer, J.: Defining virtual reality: dimensions determining telepresence. J. Commun. **42**(4), 73–93 (1992)
35. Kardong-Edgren, S.S., Farra, S.L., Alinier, G., Young, H.M.: A call to unify definitions of virtual reality. Clin. Simul. Nurs. **31**, 28–34 (2019)
36. Deutsch, D., Landsberg, P.T.: The fabric of reality. Nature **388**(6638), 136–136 (1997)
37. Brooks, F.P.: What's real about virtual reality? IEEE Comput. Graph. Appl. **19**(6), 16–27 (1999)
38. Milgram, P., Colquhoun, H., et al.: A taxonomy of real and virtual world display integration. Mixed Real. Merg. Real Virtual Worlds **1**(1999), 1–26 (1999)
39. Nincarean, D., Alia, M.B., Halim, N.D.A., Rahman, M.H.A.: Mobile augmented reality: the potential for education. Procedia-Soc. Behavior. Sci. **103**, 657–664 (2013)
40. Augmented reality: Definition of augmented reality by oxford dictionary on lexico.com also meaning of augmented reality (2024). [Online]. Available: https://www.lexico.com/en/definition/augmented_reality
41. Mixed reality (2024). [Online]. Available: https://techterms.com/definition/mixed_reality
42. Virtual Reality: Definition of virtual reality by oxford dictionary on lexico.com also meaning of virtual reality (2024). [Online]. Available: https://www.lexico.com/en/definition/virtual_reality
43. Alsop, T.: Topic: Virtual reality (VR) (2021). [Online]. Available: https://www.statista.com/topics/2532/virtual-reality-vr/#topicHeader_wrapper
44. Garner, T.A.: Applications of virtual reality. In: Echoes of Other Worlds: Sound in Virtual Reality, pp. 299–362. Springer, Berlin (2018)
45. McLellan, H.: Virtual realities. In: Handbook of Research for Educational Communications and Technology, pp. 457–487. CiteSeer (1996)
46. Cordeil, M., Dwyer, T., Klein, K., Laha, B., Marriott, K., Thomas, B.H.: Immersive collaborative analysis of network connectivity: cave-style or head-mounted display? IEEE Trans. Visualizat. Comput. Graph. **23**(1), 441–450 (2016)
47. Cave (2020). [Online]. Available: https://en.wikipedia.org/wiki/Cave_automatic_virtual_environment
48. DeFanti, T., Acevedo, D., Ainsworth, R., Brown, M., Cutchin, S., Dawe, G., Doerr, K.-U., Johnson, A., Knox, C., Kooima, R., et al.: The future of the cave. Open Eng. **1**(1), 16–37 (2011)
49. Shibata, T.: Head mounted display. Displays **23**(1–2), 57–64 (2002)
50. What is a head-mounted display (HMD)?—definition from Techopedia (2024). [Online]. Available: https://www.techopedia.com/definition/2342/head-mounted-display-hmd
51. Costello, P.J., et al.: Health and Safety Issues Associated with Virtual Reality: A Review of Current Literature. CiteSeer (1997)
52. Bowman, D.A., Datey, A., Farooq, U., Ryu, Y., Vasnaik, O.: Empirical comparisons of virtual environment displays. Virginia Polytechnic Institute & State, Technical Report, Department of Computer Science (2001)
53. Elor, A., Powell, M., Mahmoodi, E., Hawthorne, N., Teodorescu, M., Kurniawan, S.: On shooting stars: comparing cave and HMD immersive virtual reality exergaming for adults with mixed ability. ACM Trans. Comput. Healthcare **1**(4), 1–22 (2020)
54. Burdea, G., Richard, P., Coiffet, P.: Multimodal virtual reality: input-output devices, system integration, and human factors. Int. J. Human-Comput. Inter. **8**(1), 5–24 (1996)
55. Sherman, W.R., Craig, A.B.: Understanding Virtual Reality: Interface, Application, and Design. Morgan Kaufmann, Burlington (2018)

56. Broll, W., Grimm, P., Herold, R., Reiners, D., Cruz-Neira, C.: VR/AR output devices. In: Virtual and Augmented Reality (VR/AR), pp. 149–200. Springer, Berlin (2022)
57. Kim, T.-B.: Graphical menus for virtual environments. Institute for Graphical Interfaces (2005)
58. Wenjun, H.: The three-dimensional user interface. In: Advances in Human Computer Interaction. IntechOpen, London (2008)
59. Stone, D., Jarrett, C., Woodroffe, M., Minocha, S.: User Interface Design and Evaluation. Elsevier, Amsterdam (2005)
60. Bowman, D.A., Kruijff, E.: 3d User Interfaces: Theory and Practice. Addison-Wesley, Boston (2005)
61. Alger, M.: Visual design methods for virtual reality. Ravensbourne. http://aperturesciencellc.com/vr/VisualDesignMethodsforVR_MikeAlger.pdf (2015)
62. Steinicke, F., Bruder, G., Hinrichs, K., Ropinski, T., Lopes, M.: 3D user interfaces for collaborative work. In: Human Computer Interaction. IntechOpen, London (2008)
63. Stellmach, S., Dachselt, R.: Looking at 3D user interfaces. In: CHI 2012 Workshop on The 3rd Dimension of CHI (3DCHI): Touching and Designing 3D User Interfaces, pp. 95–98 (2012)
64. Marriott, K., Schreiber, F., Dwyer, T., Klein, K., Riche, N.H., Itoh, T., Stuerzlinger, W., Thomas, B.H.: Immersive Analytics, vol. 11190. Springer, Berlin (2018)
65. Zhang, X.: Virtual reality application in data visualization and analysis. M.S. Thesis, TH Köln, Köln (2017)
66. Bolton, J., Lambert, M., Lirette, D., Unsworth, B.: PaperDude: A virtual reality cycling exergame. In: CHI'14 Extended Abstracts on Human Factors in Computing Systems, pp. 475–478. ACM, New York (2014)
67. Yoo, S., Kay, J.: VRun: Running-in-place virtual reality exergame. In: Proceedings of the 28th Australian Conference on Computer-Human Interaction, pp. 562–566. ACM, New York (2016)
68. Shaw, L.A., Wünsche, B.C., Lutteroth, C., Marks, S., Callies, R.: Challenges in virtual reality exergame design. In: Proceedings of the 16th Australasian User Interface Conference (AUIC 2015), Sydney (2015)
69. Bogost, I.: The rhetoric of exergaming. In: Proceedings of the Digital Arts and Cultures (DAC) (2005)
70. Ryan, M.-L.: Immersion vs. interactivity: virtual reality and literary theory. SubStance **28**(2), 110–137 (1999)
71. Zahorik, P., Jenison, R.L.: Presence as being-in-the-world. Presence **7**(1), 78–89 (1998)
72. McGloin, R., Farrar, K., Krcmar, M.: Video games, immersion, and cognitive aggression: does the controller matter? Media psychol. **16**(1), 65–87 (2013)
73. Monahan, T., McArdle, G., Bertolotto, M.: Virtual reality for collaborative e-learning. Comput. Edu. **50**(4), 1339–1353 (2008)
74. McGrath, J.L., Taekman, J.M., Dev, P., Danforth, D.R., Mohan, D., Kman, N., Crichlow, A., Bond, W.F., Riker, S., Lemheney, A., et al.: Using virtual reality simulation environments to assess competence for emergency medicine learners. Acad. Emer. Med. **25**(2), 186–195 (2018)
75. Nee, A.Y., Ong, S.-K.: Virtual and augmented reality applications in manufacturing. IFAC Proc. Vol. **46**(9), 15–26 (2013)
76. Manresa, C., Varona, J., Mas, R., Perales, F.J.: Hand tracking and gesture recognition for human-computer interaction. ELCVIA Electr. Lett. Comput. Vis. Image Analy. **5**(3), 96–104 (2005)
77. Wang, R.Y., Popović, J.: Real-time hand-tracking with a color glove. ACM Trans. Graph. **28**(3), 1–8 (2009)
78. Taylor, J., Bordeaux, L., Cashman, T., Corish, B., Keskin, C., Sharp, T., Soto, E., Sweeney, D., Valentin, J., Luff, B., et al.: Efficient and precise interactive hand tracking through joint, continuous optimization of pose and correspondences. ACM Trans. Graph. **35**(4), 1–12 (2016)

79. Cameron, C.R., DiValentin, L.W., Manaktala, R., McElhaney, A.C., Nostrand, C.H., Quinlan, O.J., Sharpe, L.N., Slagle, A.C., Wood, C.D., Zheng, Y.Y., et al.: Hand tracking and visualization in a virtual reality simulation. In: 2011 IEEE Systems and Information Engineering Design Symposium, pp. 127–132. IEEE, Piscatway (2011)
80. McKee, M.: Biofeedback: an overview in the context of heart-brain medicine. Cleveland Clin. J. Med. **75**, S31–S34 (2008)
81. Huang, H., Wolf, S.L., He, J.: Recent developments in biofeedback for neuromotor rehabilitation. J. Neuroeng. Rehabilit. **3**(1), 11 (2006)
82. Houzangbe, S., Christmann, O., Gorisse, G., Richir, S.: Effects of voluntary heart rate control on user engagement and agency in a virtual reality game. Virtual Real. **24**(4), 665–681 (2020)
83. Michael, D.R., Chen, S.L.: Serious Games: Games That Educate, Train, and Inform. Muska & Lipman/Premier-Trade, Frederick (2005)
84. Dörner, R., Göbel, S., Effelsberg, W., Wiemeyer, J.: Serious Games. Springer, Berlin (2016)
85. Deterding, S., Sicart, M., Nacke, L., O'Hara, K., Dixon, D.: Gamification. using game-design elements in non-gaming contexts. In: CHI'11 Extended Abstracts on Human Factors in Computing Systems, pp. 2425–2428 (2011)
86. Haoran, G., Bazakidi, E., Zary, N.: Serious games in health professions education: review of trends and learning efficacy. Yearbook Med. Inf. **28**(01), 240–248 (2019)
87. Stapleton, A.J.: Serious games: Serious opportunities. In: Australian Game Developers Conference. Academic Summit, Melbourne. CiteSeer (2004)
88. Checa, D., Bustillo, A.: A review of immersive virtual reality serious games to enhance learning and training. Multimedia Tools Appl. **79**(9), 5501–5527 (2020)
89. Oh, Y., Yang, S.: Defining exergames & exergaming. In: Proceedings of Meaningful Play, pp. 1–17 (2010)
90. Cauberghe, V., De Pelsmacker, P.: Advergames. J. Advert. **39**(1), 5–18 (2010)
91. Mueller, F., Gibbs, M.R., Vetere, F.: Design influence on social play in distributed exertion games. In: Proceedings of the SIGCHI Conference on Human Factors in Computing Systems, ACM, pp. 1539–1548 (2009)
92. Murray, E.G., Neumann, D.L., Moffitt, R.L., Thomas, P.R.: The effects of the presence of others during a rowing exercise in a virtual reality environment. Psychol. Sport Exer. **22**, 328–336 (2016)
93. Demers, M., Martinie, O., Winstein, C., Robert, M.T.: Active video games and low-cost virtual reality: an ideal therapeutic modality for children with physical disabilities during a global pandemic. Front. Neurol. **11**, 1737 (2020)
94. Corregidor-Sánchez, A.I., Polonio-López, B., Martin-Conty, J.L., Rodríguez-Hernández, M., Mordillo-Mateos, L., Schez-Sobrino, S., Criado-Álvarez, J.J.: Exergames to prevent the secondary functional deterioration of older adults during hospitalization and isolation periods during the covid-19 pandemic. Sustainability **13**(14), 7932 (2021)
95. Brown, C.M.: Human-Computer Interface Design Guidelines. Intellect Books, Bristol (1999)
96. Hassenzahl, M., Tractinsky, N.: User experience-a research agenda. Behav. Inf. Technol. **25**(2), 91–97 (2006). https://doi.org/10.1080/01449290500330331
97. Hartson, R., Pyla, P.S.: The UX Book: Process and Guidelines for Ensuring a Quality User Experience. Elsevier, Amsterdam (2012)
98. Kaptelinin, V., Nardi, B., Bødker, S., Carroll, J., Hollan, J., Hutchins, E., Winograd, T.: Post-cognitivist HCI: Second-wave theories. In: CHI'03 Extended Abstracts on Human Factors in Computing Systems, pp. 692–693 (2003)
99. Calvo, R.A., D'Mello, S., Gratch, J.M., Kappas, A.: The Oxford Handbook of Affective Computing. Oxford Library of Psychology, Oxford (2015)
100. Norman, D., Miller, J., Henderson, A.: What you see, some of what's in the future, and how we go about doing it: Hi at apple computer. In: Conference Companion on Human Factors in Computing Systems, p. 155 (1995)
101. Roto, V., Law, E., Vermeeren, A., Hoonhout, J.: User experience white paper. Bringing clarity to the concept of user experience. Result from Dagstuhl seminar on demarcating user experience, September 15–18 (2010). Disponible en ligne le **22**, 06–15 (2011)

102. Wechsung, I., Moor, K.D.: Quality of experience versus user experience. In: Quality of Experience, pp. 35–54. Springer, Berlin (2014)
103. Wilson, G.M., Angela Sasse, M.: From doing to being: getting closer to the user experience. Interact. Comput. **16**(4), 697–705 (2004)
104. Qualinet White Paper on Definitions of Quality of Experience. COST Action IC 1003, Le Callet, P., Möller, S., Perkis, A. (Eds.) (2013)
105. Jekosch, U.: Voice and Speech Quality Perception: Assessment and Evaluation. Springer, Berlin (2005). https://doi.org/10.1007/3-540-28860-0
106. Bargas-Avila, J.A., Hornbæk, K.: Old Wine in New Bottles or Novel Challenges: A Critical Analysis of Empirical Studies of User Experience. Association for Computing Machinery, New York (2011). ISBN: 978-14-50302-28-9. https://doi.org/10.1145/1978942.1979336
107. Tcha-Tokey, K., Christmann, O., Loup-Escande, E., Loup, G., Richir, S.: Towards a model of user experience in immersive virtual environments. In: Advances in Human-Computer Interaction, vol. 2018 (2018)
108. Heater, C.: Being there: the subjective experience of presence. Presence Teleoper. Virtual Environ. **1**(2), 262–271 (1992)
109. Bulu, S.T.: Place presence, social presence, co-presence, and satisfaction in virtual worlds. Comput. Edu. **58**(1), 154–161 (2012)
110. Bowman, D.A., McMahan, R.P.: Virtual reality: How much immersion is enough? Computer **40**(7), 36–43 (2007)
111. McCall, R., O'Neil, S., Carroll, F.: Measuring presence in virtual environments. In: CHI'04 Extended Abstracts on Human Factors in Computing Systems, pp. 783–784 (2004)
112. Csikszentmihalyi, M.: Beyond Boredom and Anxiety. Jossey-Bass, Hoboken (2000)
113. Geslin, E., Bouchard, S., Richir, S.: Gamers' versus non-gamers' emotional response in virtual reality. J. CyberTher. Rehabil. **4**(4), 489–493 (2011)
114. Cheng, L.-K., Chieng, M.-H., Chieng, W.-H.: Measuring virtual experience in a three-dimensional virtual reality interactive simulator environment: a structural equation modeling approach. Virtual Real. **18**(3), 173–188 (2014)
115. Albert, B., Tullis, T.: Measuring the User Experience: Collecting, Analyzing, and Presenting Usability Metrics. Newness, London (2013)
116. Card, S.K., Moran, T.P., Newell, A.: The Psychology of Human-Computer Interaction. CRC Press, Boca Raton (2018)
117. Lund, A.M.: Measuring usability with the use questionnaire12. Usabil. Inter. **8**(2), 3–6 (2001)
118. Bangor, A., Kortum, P.T., Miller, J.T.: An empirical evaluation of the system usability scale. Int. J. Human-Comput. Inter. **24**(6), 574–594 (2008)
119. Brooke, J., et al.: SUS-a quick and dirty usability scale. Usabil. Evaluat. Ind. **189**(194), 4–7 (1996)
120. A. S. for Public Affairs: System Usability Scale (SUS). Department of Health and Human Services 2013 (2013).
121. Bradley, M.M., Lang, P.J.: Measuring emotion: the self-assessment manikin and the semantic differential. J. Behav. Therapy Exp. Psych. **25**(1), 49–59 (1994)
122. Hassenzahl, M., Trautmann, T.: Analysis of web sites with the repertory grid technique. In: CHI'01 Extended Abstracts on Human Factors in Computing Systems, pp. 167–168 (2001)
123. Hassenzahl, M., Burmester, M., Koller, F.: Attrakdiff: Ein fragebogen zur messung wahrgenommener hedonischer und pragmatischer qualität. In: Mensch & Computer 2003, pp. 187–196. Springer, Berlin (2003)
124. Laugwitz, B., Held, T., Schrepp, M.: Construction and evaluation of a user experience questionnaire. In: Symposium of the Austrian HCI and Usability Engineering Group, pp. 63–76. Springer, Berlin (2008)
125. Schrepp, M., Hinderks, A., Thomaschewski, J.: Design and evaluation of a short version of the user experience questionnaire (UEQ-S). IJIMAI **4**(6), 103–108 (2017)
126. Díaz-Oreiro, I., López, G., Quesada, L., Guerrero, L.A.: Standardized questionnaires for user experience evaluation: a systematic literature review. Multidiscipl. Digital Publ. Instit. Proc. **31**(1), 14 (2019)

127. Boyle, E.A., Connolly, T.M., Hainey, T., Boyle, J.M.: Engagement in digital entertainment games: a systematic review. Comput. Human Behav. **28**(3), 771–780 (2012)
128. Huang, W., Roscoe, R.D., Johnson-Glenberg, M.C., Craig, S.D.: Motivation, engagement, and performance across multiple virtual reality sessions and levels of immersion. J. Comput. Assist. Learn. **37**(3), 745–758 (2021)
129. Keller, J.M.: Motivation, learning, and technology: applying the arcs-v motivation model. Participat. Edu. Res. **3**(2), 1–15 (2016)
130. Dalgarno, B., Lee, M.J.: What are the learning affordances of 3-d virtual environments? Brit. J. Edu. Technol. **41**(1), 10–32 (2010)
131. Chi, M.T., Wylie, R.: The ICAP framework: linking cognitive engagement to active learning outcomes. Edu. Psychol. **49**(4), 219–243 (2014)
132. Ryan, R.M., Deci, E.L.: Self-determination theory and the facilitation of intrinsic motivation, social development, and well-being. Am. Psychol. **55**(1), 68 (2000)
133. Ryan, R.M., Rigby, C.S., Przybylski, A.: The motivational pull of video games: a self-determination theory approach. Motivat. Emot. **30**(4), 344–360 (2006)
134. Pelletier, L.G., Rocchi, M.A., Vallerand, R.J., Deci, E.L., Ryan, R.M.: Validation of the revised sport motivation scale (SMS-II). Psychol. Sport Exer. **14**(3), 329–341 (2013)
135. Richard, M., Christina, M.F., Deborah, L.S., Rubio, N., Kennon, M.S.: Intrinsic motivation and exercise adherence. Int. J. Sport. Psychol. **28**(4), 335–354 (1997)
136. Kiili, K., Perttula, A., Lindstedt, A., Arnab, S., Suominen, M.: Flow experience as a quality measure in evaluating physically activating collaborative serious games. Int. J. Ser. Games **1**(3), 35–49 (2014)
137. Möller, S., Raake, A.: Quality of Experience: Advanced Concepts, Applications and Methods. Springer, Berlin (2014)
138. Sweetser, P., Johnson, D., Wyeth, P., Anwar, A., Meng, Y., Ozdowska, A.: GameFlow in different game genres and platforms. Comput. Entertain. **15**(3), 1–24 (2017)
139. Jackson, S.A., Eklund, R.C.: Assessing flow in physical activity: the flow state scale-2 and dispositional flow scale-2. J. Sport Exer. Psychol. **24**(2), 133–150 (2002)
140. Slater, M.: Place illusion and plausibility can lead to realistic behaviour in immersive virtual environments. Philosoph. Trans. R. Soc. B Biol. Sci. **364**(1535), 3549–3557 (2009)
141. Lee, K.M.: Presence, explicated. Commun. Theory **14**(1), 27–50 (2004)
142. Aymerich-Franch, L., Karutz, C., Bailenson, J.N., et al.: Effects of facial and voice similarity on presence in a public speaking virtual environment. In: Proceedings of the International Society for Presence Research Annual Conference, pp. 24–26 (2012)
143. Biocca, F., Harms, C., Burgoon, J.K.: Toward a more robust theory and measure of social presence: review and suggested criteria. Presence: Teleop. Virtual Environ. **12**(5), 456–480 (2003)
144. Ratan, R.A., Hasler, B.: Self-presence standardized: Introducing the self-presence questionnaire (SPQ). In: Proceedings of the 12th Annual International Workshop on Presence, vol. 81 (2009)
145. Short, J., Williams, E., Christie, B.: The Social Psychology of Telecommunications. Wiley, Toronto (1976)
146. Lee, K.M., Jung, Y., Kim, J., Kim, S.R.: Are physically embodied social agents better than disembodied social agents?: The effects of physical embodiment, tactile interaction, and people's loneliness in human-robot interaction. Int. J. Human-Comput. Stud. **64**(10), 962–973 (2006)
147. Grassini, S., Laumann, K.: Questionnaire measures and physiological correlates of presence: a systematic review. Front. Psychol. **11**, 349 (2020)
148. Witmer, B.G., Singer, M.J.: Measuring presence in virtual environments: a presence questionnaire. Presence **7**(3), 225–240 (1998)
149. Schubert, T., Friedmann, F., Regenbrecht, H.: The experience of presence: factor analytic insights. Presence **10**(3), 266–281 (2001)
150. Kober, S.E., Neuper, C.: Personality and presence in virtual reality: does their relationship depend on the used presence measure? Int. J. Human-Comput. Inter. **29**(1), 13–25 (2013)

151. Szczurowski, K., Smith, M.: Measuring presence: Hypothetical quantitative framework. In: 2017 23rd International Conference on Virtual System & Multimedia (VSMM), pp. 1–8. IEEE, Piscatway (2017)
152. Oh, C.S., Bailenson, J.N., Welch, G.F.: A systematic review of social presence: definition, antecedents, and implications. Front. Robot. AI **5** , 114 (2018)
153. Biocca, F., Harms, C.: Defining and measuring social presence: contribution to the networked minds theory and measure. Proc. PRESENCE **2002**, 1–36 (2002)
154. Nesbitt, K., Nalivaiko, E.: Cybersickness. In: Encyclopedia of Computer Graphics and Games, Lee, N. (Ed.), pp. 1–6. Springer International Publishing, Cham (2018). ISBN: 978-3-319-08234-9. https://doi.org/10.1007/978-3-319-08234-9_252-1
155. Diels, C., Bos, J.E.: Self-driving carsickness. Appl. Ergon. **53**, 374–382 (2016)
156. Gavgani, A.M., Hodgson, D.M., Nalivaiko, E.: Effects of visual flow direction on signs and symptoms of cybersickness. PLoS one **12**(8), e0182790 (2017)
157. Classen, S., Bewernitz, M., Shechtman, O.: Driving simulator sickness: an evidence-based review of the literature. Am. J. Occupat. Therapy **65**(2), 179–188 (2011)
158. Waltemate, T., Senna, I., Hülsmann, F., Rohde, M., Kopp, S., Ernst, M., Botsch, M.: The impact of latency on perceptual judgments and motor performance in closed-loop interaction in virtual reality. In: Proceedings of the 22nd ACM Conference on Virtual Reality Software and Technology, pp. 27–35. ACM, New York (2016)
159. Jörg, S., Normoyle, A., Safonova, A.: How responsiveness affects players' perception in digital games. In: Proceedings of the ACM Symposium on Applied Perception, pp. 33–38. ACM, New York (2012)
160. Meehan, M., Razzaque, S., Whitton, M.C., Brooks, F.P.: Effect of latency on presence in stressful virtual environments. In: IEEE Virtual Reality, 2003 Proceedings, pp. 141–148. IEEE, Piscataway (2003)
161. Roufs, J.A., Boschman, M.C.: Text quality metrics for visual display units: I. methodological aspects. Displays **18**(1), 37–43 (1997)
162. Ardoin, S.P., Suldo, S.M., Witt, J., Aldrich, S., McDonald, E.: Accuracy of readability estimates' predictions of CBM performance. School Psychol. Quart. **20**(1), 1 (2005)
163. Pikulski, J.J.: Readability. Retrieved September, vol. 20, p. 2014 (2002)
164. Ali, A.Z.M., Wahid, R., Samsudin, K., Idris, M.Z.: Reading on the computer screen: does font type have effects on web text readability? Int. Edu. Stud. **6**(3), 26–35 (2013)
165. Developers, G.: Designing Screen Interfaces for VR (Google i/o'17) (2017)
166. Grout, C., Rogers, W., Apperley, M., Jones, S.: Reading text in an immersive head-mounted display: An investigation into displaying desktop interfaces in a 3d virtual environment. In: Proceedings of the 15th New Zealand Conference on Human-Computer Interaction, pp. 9–16. ACM, New York (2015)
167. Dingler, T., Kunze, K., Outram, B.: VR reading UIS: Assessing text parameters for reading in VR. In: Extended Abstracts of the 2018 CHI Conference on Human Factors in Computing Systems, LBW094. ACM, New York (2018)
168. Hall, R.H., Hanna, P.: The impact of web page text-background colour combinations on readability, retention, aesthetics and behavioural intention. Behav. Inf. Technol. **23**(3), 183–195 (2004)
169. Yates, R.: Web site accessibility and usability: Towards more functional sites for all. Campus-Wide Inf. Syst. **22**, 180–188 (2005)
170. Web content accessibility guidelines (WCAG) 2.1 (2024). [Online]. Available: https://www.w3.org/TR/WCAG21/#abstract
171. Distler, V., Lallemand, C., Bellet, T.: Acceptability and acceptance of autonomous mobility on demand: The impact of an immersive experience. In: Proceedings of the 2018 CHI Conference on Human Factors in Computing Systems, pp. 1–10 (2018)
172. Eghbali, P.: Social acceptability of virtual reality interaction: Experiential factors and design implications (2018)

173. Schwind, V., Reinhardt, J., Rzayev, R., Henze, N., Wolf, K.: Virtual reality on the go? a study on social acceptance of VR glasses. In: Proceedings of the 20th International Conference on Human-Computer Interaction with Mobile Devices and Services Adjunct, pp. 111–118 (2018)
174. Mai, C., Wiltzius, T., Alt, F., Hußmann, H.: Feeling alone in public: Investigating the influence of spatial layout on users' VR experience. In: Proceedings of the 10th Nordic Conference on Human-Computer Interaction, pp. 286–298 (2018)
175. Ahsanullah, S.S., Kamil, M., Muzafar, K.: Understanding factors influencing user experience of interactive systems: a literature review. ARPN J. Eng. Appl. Sci. **10**(18) 175–185 (2006)
176. Bevan, N.: Classifying and selecting UX and usability measures. In: International Workshop on Meaningful Measures: Valid Useful User Experience Measurement, vol. 11, pp. 13–18 (2008)
177. Forlizzi, J., Battarbee, K.: Understanding experience in interactive systems. In: Proceedings of the 5th Conference on Designing Interactive Systems: Processes, Practices, Methods, and Techniques, pp. 261–268 (2004)
178. Adikari, S., McDonald, C., Campbell, J.: A design science framework for designing and assessing user experience. In: International Conference on Human-Computer Interaction, pp. 25–34. Springer, Berlin (2011)
179. Law, E.L.-C., Sun, X.: Evaluating user experience of adaptive digital educational games with activity theory. Int. J. Human-Comput. Stud. **70**(7), 478–497 (2012)
180. Le Callet, P., Möller, S., Perkis, A., et al.: Qualinet white paper on definitions of quality of experience. In: European Network on Quality of Experience in Multimedia Systems and Services (COST Action IC 1003), vol. 3 (2012)
181. Wienrich, C., Döllinger, N., Kock, S., Schindler, K., Traupe, O.: Assessing user experience in virtual reality–a comparison of different measurements. In: International Conference of Design, User Experience, and Usability, pp. 573–589. Springer, Berlin (2018)
182. Zarour, M., Alharbi, M.: User experience framework that combines aspects, dimensions, and measurement methods. Cogent Eng. **4**(1), 1421006 (2017)
183. Vermeeren, A.P., Law, E.L.-C., Roto, V., Obrist, M., Hoonhout, J., Väänänen-Vainio-Mattila, K. : User experience evaluation methods: Current state and development needs. In: Proceedings of the 6th Nordic Conference on Human-Computer Interaction: Extending Boundaries, pp. 521–530 (2010)
184. Kim, Y.M., Rhiu, I., Yun, M.H.: A systematic review of a virtual reality system from the perspective of user experience. Int. J. Human-Comput. Interact. **36**(10), 893–910 (2020)
185. González Sánchez, J.L., Padilla Zea, N., Gutiérrez, F.L.: From usability to playability: Introduction to player-centred video game development process. In: International Conference on Human Centered Design, pp. 65–74. Springer, Berlin (2009)
186. Möller, S., Schmidt, S., Zadtootaghaj, S.: New ITU-T standards for gaming QoE evaluation and management. In: 2018 Tenth International Conference on Quality of Multimedia Experience (QoMEX), pp. 1–6. IEEE, Piscataway (2018). https://doi.org/10.1109/QoMEX.2018.8463404
187. Nacke, L.E., Drachen, A., Göbel, S.: Methods for evaluating gameplay experience in a serious gaming context. Int. J. Comput. Sci. Sport **9**(2), 1–12 (2010)
188. Le Marc, C., Mathieu, J.-P., Pallot, M., Richir, S.: Serious gaming: From learning experience towards user experience. In: 2010 IEEE International Technology Management Conference (ICE), pp. 1–12. IEEE, Piscataway (2010)
189. ITU-T Recommendation G.1032: Influence Factors on Gaming Quality of Experience. International Telecommunication Union, Geneva (2017)
190. Sutcliffe, A., Gault, B., Fernando, T., Tan, K.: Investigating interaction in CAVE virtual environments. ACM Trans. Comput.-Human Inter. **13**(2), 235–267 (2006)
191. Juan, M.C., Pérez, D.: Comparison of the levels of presence and anxiety in an acrophobic environment viewed via HMD or CAVE. Presence Teleop. Virtual Environ. **18**(3), 232–248 (2009)

192. Fonseca, D., Kraus, M.: A comparison of head-mounted and hand-held displays for 360°videos with focus on attitude and behavior change. In: Proceedings of the 20th International Academic Mindtrek Conference, pp. 287–296. ACM, New York (2016)
193. Peng, W., Lin, J.-H., Pfeiffer, K.A., Winn, B.: Need satisfaction supportive game features as motivational determinants: an experimental study of a self-determination theory guided exergame. Media Psychol. **15**(2), 175–196 (2012)
194. Mouatt, B., Smith, A.E., Mellow, M.L., Parfitt, G., Smith, R.T., Stanton, T.R.: The use of virtual reality to influence motivation, affect, enjoyment, and engagement during exercise: a scoping review. Front. Virtual Real. **1**, 564–664 (2020)
195. Mania, K., Chalmers, A.: The effects of levels of immersion on memory and presence in virtual environments: a reality centered approach. CyberPsychol. Behavior **4**(2), 247–264 (2001)
196. Penelle, B., Debeir, O.: Multi-sensor data fusion for hand tracking using Kinect and Leap Motion. In: Proceedings of the 2014 Virtual Reality International Conference, pp. 1–7 (2014)
197. Zyda, M.: From visual simulation to virtual reality to games. Computer **38**(9), 25–32 (2005)
198. Alallah, F., Neshati, A., Sakamoto, Y., Hasan, K., Lank, E., Bunt, A., Irani, P.: Performer vs. observer: Whose comfort level should we consider when examining the social acceptability of input modalities for head-worn display? In: Proceedings of the 24th ACM Symposium on Virtual Reality Software and Technology, pp. 1–9 (2018)
199. Eghbali, P., Väänänen, K., Jokela, T.: Social acceptability of virtual reality in public spaces: Experiential factors and design recommendations. In: Proceedings of the 18th International Conference on Mobile and Ubiquitous Multimedia, pp. 1–11 (2019)
200. Valentini, I., Ballestin, G., Bassano, C., Solari, F., Chessa, M.: Improving obstacle awareness to enhance interaction in virtual reality. In: 2020 IEEE Conference on Virtual Reality and 3D User Interfaces (VR), pp. 44–52. IEEE, Piscataway (2020)
201. Wehbe, R.R., Nacke, L.E.: Towards understanding the importance of co-located gameplay. In: Proceedings of the 2015 Annual Symposium on Computer-Human Interaction in Play, pp. 733–738. ACM, New York (2015)
202. Burgoon, J.K., Bonito, J.A., Ramirez, A., Dunbar, N.E., Kam, K., Fischer, J.: Testing the interactivity principle: effects of mediation, propinquity, and verbal and nonverbal modalities in interpersonal interaction. J. Commun. **52**(3), 657–677 (2002)
203. Skowronek, J., Raake, A., Berndtsson, G., Rummukainen, O.S., Usai, P., Gunkel, S.N., Johanson, M., Habets, E.A., Malfait, L., Lindero, D., et al.: Quality of experience in telemeetings and videoconferencing: A comprehensive survey. IEEE Access **10**, 63885–63931 (2022)
204. Comber, C., Colley, A., Hargreaves, D.J., Dorn, L.: The effects of age, gender and computer experience upon computer attitudes. Edu. Res. **39**(2), 123–133 (1997)
205. Roberts, S.C., Mehler, B., Orszulak, J., Reimer, B., Coughlin, J., Glass, J.: An evaluation of age, gender, and technology experience in user performance and impressions of a multimodal human-machine interface. In: IIE Annual Conference. Proceedings, Institute of Industrial and Systems Engineers (IISE), p. 1 (2011)
206. Thayer, S.E., Ray, S.: Online communication preferences across age, gender, and duration of internet use. CyberPsychol. Behavior **9**(4), 432–440 (2006)
207. Narciso, D., Bessa, M., Melo, M., Coelho, A., Vasconcelos-Raposo, J.: Immersive 360 video user experience: impact of different variables in the sense of presence and cybersickness. Univer. Access Inf. Soc. **18**(1), 77–87 (2019)
208. Petri, K., Feuerstein, K., Folster, S., Bariszlovich, F., Witte, K.: Effects of age, gender, familiarity with the content, and exposure time on cybersickness in immersive head-mounted display based virtual reality. Am. J. Biomed. Sci. **12**(2), 107–112 (2020)
209. Lacko, J.: Health safety training for industry in virtual reality. In: 2020 Cybernetics & Informatics (K&I), pp. 1–5. IEEE, Piscataway (2020)
210. Molina, K.I., Ricci, N.A., de Moraes, S.A., Perracini, M.R.: Virtual reality using games for improving physical functioning in older adults: a systematic review. J. Neuroeng. Rehabilit. **11**(1), 1–20 (2014)

211. Riva, G.: Virtual reality for health care: the status of research. Cyberpsychol. Behavior **5**(3), 219–225 (2002)
212. Lu, A.S., Kharrazi, H.: A state-of-the-art systematic content analysis of games for health. Games Health J. **7**(1), 1–15 (2018)
213. Messenger, J.C., Gschwind, L.: Three generations of telework: new ICTs and the (R)evolution from home office to virtual office. New Technol. Work Empl. **31**(3), 195–208 (2016)
214. Kojo, I., Nenonen, S.: Evolution of co-working places: drivers and possibilities. Intell. Buil. Int. **9**(3), 164–175 (2017)
215. Bowman, B., Elmqvist, N., Jankun-Kelly, T.: Toward visualization for games: theory, design space, and patterns. IEEE Trans. Visualiz. Comput. Graph. **18**(11), 1956–1968 (2012)
216. Llanos, S.C., Jørgensen, K.: Do players prefer integrated user interfaces? A qualitative study of game UI design issues. In: Proceedings of DiGRA 2011 Conference: Think Design Play (2011)
217. Vatavu, R.-D., Ungurean, O.-C., Pentiuc, S.-G.: Body gestures for office desk scenarios. In: Whole Body Interaction, pp. 163–172. Springer, Berlin (2011)
218. Millais, P., Jones, S.L., Kelly, R.: Exploring data in virtual reality: Comparisons with 2d data visualizations. In: Extended Abstracts of the 2018 CHI Conference on Human Factors in Computing Systems, pp. 1–6 (2018)
219. Virtual reality is a big trend in museums, but what are the best examples of museums using VR? (2022)
220. Guttentag, D.A.: Virtual reality: Applications and implications for tourism. Tourism Manag. **31**(5), 637–651 (2010)
221. Sygic Travel (2024). [Online]. Available: https://www.youtube.com/channel/UCN027-rS7Z7QmmR336HJA1Q
222. Sygic Travel Maps (2024). [Online]. Available: https://www.sygic.com/it/travel
223. Jackson, S.A., Eklund, R.C.: The flow scales manual. Fitness Information Technology (2004)
224. Schubert, T., Friedmann, F., Regenbrecht, H.: Embodied presence in virtual environments. In: Visual Representations and Interpretations, pp. 269–278. Springer, Berlin (1999)
225. Biocca, F., Harms, C., Gregg, J.: The networked minds measure of social presence: Pilot test of the factor structure and concurrent validity. In: 4th Annual International Workshop on Presence, Philadelphia, pp. 1–9 (2001)
226. Jackson, S.A., Martin, A.J., Eklund, R.C.: Long and short measures of flow: the construct validity of the fss-2, dfs-2, and new brief counterparts. J. Sport Exer. Psychol. **30**(5), 561–587 (2008)
227. Jackson, S.A., Marsh, H.W.: Development and validation of a scale to measure optimal experience: the flow state scale. J. Sport Exer. Psychol. **18**(1), 17–35 (1996)
228. Hattie, J., Timperley, H.: The power of feedback. Rev. Educ. Res. **77**(1), 81–112 (2007)
229. Meuret, A.E., Wilhelm, F.H., Ritz, T., Roth, W.T.: Breathing training for treating panic disorder: useful intervention or impediment? Behavior Modif. **27**(5), 731–754 (2003)
230. Smith, R.M., Loschner, C.: Biomechanics feedback for rowing. J. Sports Sci. **20**(10), 783–791 (2002)
231. Karim, N.S.A., Hasan, A.: Reading habits and attitude in the digital age. The Electronic Library (2007)
232. Caldwell, B., Cooper, M., Reid, L.G., Vanderheiden, G.: Web content accessibility guidelines (WCAG) 2.0. WWW Consortium (W3C) (2008)
233. Franke, T., Attig, C., Wessel, D.: A personal resource for technology interaction: development and validation of the affinity for technology interaction (ATI) scale. Int. J. Human-Comput. Inter. **35**(6), 456–467 (2019)
234. Schubert, T., Friedmann, F., Regenbrecht, H.: The experience of presence: factor analytic insights. Presence: Teleop. Virtual Environ. **10**(3), 266–281 (2001)
235. Slater, M., Wilbur, S.: A framework for immersive virtual environments (five): speculations on the role of presence in virtual environments. Presence Teleop. Virtual Environ. **6**(6), 603–616 (1997)

236. Crotty, M.: The Foundations of Social Research: Meaning and Perspective in the Research Process. Sage, Thousand Oaks (1998)
237. Blanca, M.J., Alarcón, R., Arnau, J., Bono, R., Bendayan, R.: Non-normal data: is ANOVA still a valid option? Psicothema **29**(4), 552–557 (2017)
238. Ryan, R.M., Rigby, C.S., Przybylski, A.: The motivational pull of video games: a self-determination theory approach. Motivat. Emot. **30**(4), 344–360 (2006)
239. Weibel, D., Wissmath, B., Habegger, S., Steiner, Y., Groner, R.: Playing online games against computer-vs. human-controlled opponents: effects on presence, flow, and enjoyment. Comput. Human Behav. **24**(5), 2274–2291 (2008)
240. Schmidt, S., Zadtootaghaj, S., Möller, S.: Towards the delay sensitivity of games: There is more than genres. In: 2017 Ninth International Conference on Quality of Multimedia Experience (QoMEX), pp. 1–6. IEEE, Piscataway (2017). https://doi.org/10.1109/QoMEX.2017.7965676
241. Mousas, C., Kao, D., Koilias, A., Rekabdar, B.: Real and virtual environment mismatching induces arousal and alters movement behavior. In: 2020 IEEE Conference on Virtual Reality and 3D User Interfaces (VR), pp. 626–635. IEEE, Piscataway (2020)

Index

T. Kojić, *User Experience for Serious Games in Virtual Reality*, T-Labs Series in Telecommunication Services, https://doi.org/10.1007/978-3-031-75530-9

MIX
Papier aus verantwortungsvollen Quellen
Paper from responsible sources
FSC® C105338

If you have any concerns about our products,
you can contact us on
ProductSafety@springernature.com

In case Publisher is established outside the EU,
the EU authorized representative is:
Springer Nature Customer Service Center GmbH
Europaplatz 3, 69115 Heidelberg, Germany

Printed by Libri Plureos GmbH
in Hamburg, Germany